W9-AGD-285

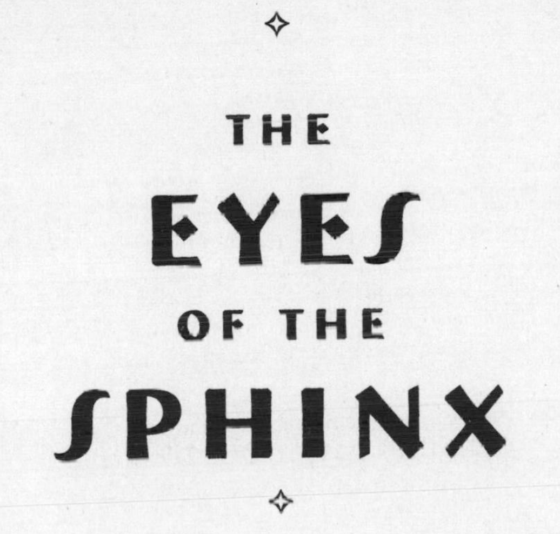
THE
EYES
OF THE
SPHINX

Berkley Books by Erich von Däniken

THE EYES OF THE SPHINX
CHARIOTS OF THE GODS
SIGNS OF THE GODS
PATHWAYS TO THE GODS

THE EYES OF THE SPHINX

THE NEWEST EVIDENCE OF EXTRATERRESTRIAL CONTACT IN ANCIENT EGYPT

Erich von Däniken

THE EYES OF THE SPHINX

A Berkley Book / published by arrangement with
the author

PRINTING HISTORY
Berkley trade paperback edition / March 1996

Book design by Stanley S. Drate / Folio Graphics, Inc.

For information address:
The Berkley Publishing Group, a division of Penguin Putnam Inc.,
375 Hudson Street, New York, New York 10014.

The Penguin Putnam Inc. World Wide Web site address is
http://www.penguinputnam.com

ISBN: 0-425-15130-1

BERKLEY®
Berkley Books are published by
The Berkley Publishing Group, a division of Penguin Putnam Inc.,
375 Hudson Street, New York, New York 10014.
BERKLEY and the "B" design are trademarks
belonging to Penguin Putnam Inc.

PRINTED IN THE UNITED STATES OF AMERICA

15

Contents

THE EYES OF THE SPHINX

✧ 1 ✧

Pet Cemeteries and Empty Tombs

"O Egypt! Egypt! Your knowledge will survive but in legends, which later generations will be unable to believe."

LUCIUS APULEIUS,
ROMAN PHILOSOPHER,
SECOND CENTURY A.D.

"Welcome to Egypt!" A lanky young man with a black mustache stepped in my way and extended his hand in greeting. Somewhat surprised, I grabbed it, thinking to myself that this had to be the newest way to welcome tourists to the country. Immediately, I was subjected to the typical questioning: Where was I from and what did I intend to visit in Egypt? Jovially, though somewhat awkwardly, I got rid of this pushy fellow. But not for long. I had barely left the confines of the Cairo airport building when I was confronted by another one just like him: "Welcome to Egypt!" My suitcases, please. More semivoluntary handshaking.

In the days that followed, the same bothersome treatment repeated itself numerous times. "Welcome to Egypt!" rang from the steps of the Egyptian Antiquities Museum in Cairo. "Welcome to Egypt!" proclaimed the papyrus seller. "Welcome to Egypt!" from the shoeshine boy at the street corner, the taxi driver, the hotel receptionist, and the clerk at the souvenir store.

Everyone wanted to know what country I was from, and I was getting tired of answering the same question over and over. Having my hand shaken and my origins inquired after for the forty-second time in front of the step pyramid of Saqqâra, I answered solemnly, "I am from Mars." Undaunted by my reply, the man grabbed both of my hands and declared emphatically once more: "Welcome to Egypt!"

That's how far the Egyptians have come: Even tourists from Mars don't seem to be anything new!

I have traveled to this country on the Nile many times in my sixty years. Many things have changed—the streets, the transportation, the polluted air, the new hotel complexes—but a shroud of mystery continues to linger over it. It still generates the same reverent fascination for which Egypt has always been known.

Back in 1954, as a nineteen-year-old, I first descended into the subterranean tunnels under the desert sand of Saqqâra. An Egyptian friend and two guards led the way. Each of us carried burning candles, because forty years ago the stuffy vaults were without electricity. The tunnels had not yet been opened to tourists. I remember very clearly how one of the guards raised his candle to highlight a massive, man-sized sarcophagus. The trembling flame reflected weakly off the granite block.

"What's in it?" I asked hesitantly.

"Sacred bulls, young man, mummified bulls!"

A few feet away was another wide recess in the vault with another bull sarcophagus, and yet another one across from it. Giant sarcophaguses lined the musty vaults as far as the candlelight could reach. A thick carpet of dust muffled our steps like plush velvet. More corridors, more recesses, more sarcophaguses. It felt rather spooky. The microscopic dust was irritating my throat. There was no breeze to circulate the stale, heavy air. All the coffins had been opened. The massive granite covers had been pushed aside slightly. I was eager to see one of the mummies and asked my friend and the two guards for assistance. They raised me up until I could lie flat on the upper edge of one of the sarcophaguses. With my candle I lit the inside of the coffin. It was squeaky clean—and empty! I tried four more, with the same result. Where were the mummified bulls? Had the huge animal corpses been removed? Were these sacred mummies now in museums? Vaguely, the suspicion arose in me that the sarcophaguses might never have held any mummified bulls.

Now, over forty years later, I was once again visiting the underground vaults. Electric lights have been installed; tourists are channeled through two parallel hallways. Many of the tourists gasp in awe as the travel guide explains knowingly that each of the huge sarcophaguses used to house the mummy of a sacred Apis bull.

I don't feel like contradicting the guide even though I know better now: No mummies of sacred bulls have ever been found in the giant granite sarcophaguses.

It Started with Auguste Mariette

Paris 1850. Twenty-eight-year-old Auguste Mariette was employed as a science assistant at the Louvre. A short, lively man who could curse like a drunken sailor, he had accumulated a wealth of knowledge about Egypt in the past seven years. He was fluent in English, French, and Arabic, was proficient at deciphering hieroglyphs, and was busy at work translating ancient Egyptian texts. The French had heard that the British, their fiercest archaeological rivals, were buying ancient texts in Egypt. Naturally, the French, in their grand tradition, could not allow themselves to be left behind. The Academy of Sciences in Paris decided to dispatch science assistant Auguste Mariette to Egypt. With six thousand francs in his pockets, he was supposed to snatch the best papyri from the British.

Auguste Mariette arrived in Cairo on October 2, 1850. The very next day, he paid a visit to the Coptic patriarchate, hoping to secure some ancient Egyptian papyrus texts from Coptic monasteries. While strolling through the antique shops of Cairo, he noticed that every store was selling genuine sphinxes, all of them originally from Saqqâra. This started Mariette thinking. When the Coptic patriarchate informed him on October 17 that more time was needed to decide the fate of the ancient papyri, Mariette was disappointed. He climbed to the top of the citadel, where he sat on the steps for a long time, pondering his options.

Below him, Cairo was shrouded in an early-evening mist. Mariette wrote that three hundred minarets protruded from the dense sea of fog like the masts of a sunken fleet. In the west, he saw the pyramids bathed in the golden flame dust of the setting sun. It was an overwhelming sight. Mariette described how captivated he was, mesmerized by an almost painful fascination. It was the fulfillment of his lifelong dream. Over there, virtually within reach, lay an entire world of tombs, monuments, inscriptions, and statues. What else was there? The next day, he rented two or three mules for his luggage and a donkey or two for himself. He bought a tent and a few cases of provisions suitable for desert travel, and on October 20, 1850, he set up his tent at the foot of the Great Pyramid.[1]

Seven days later, Mariette, impatient by nature, was fed up with the hustle and bustle around the pyramids. He moved his small caravan half a day's journey south and set up camp in Saqqâra among ruins and toppled pillars. Saqqâra's modern-day trademark, the step pyramid dedicated to the pharaoh Djoser (2630–2611 B.C.), was still hidden beneath the desert sands. Idleness was not in Auguste Mariette's character. While exploring the surroundings, he discovered the head of a sphinx protruding from the sand. Immediately, he remembered the sphinxes in the antique stores, which came from Saqqâra. A few feet away, he stumbled across a broken stone tablet, from which he deciphered the word "apis." That was enough to jolt the attention of the young man from Paris. Other visitors before Auguste Mariette had seen the sphinx head and the stone tablet, but nobody had made the connection. Mariette recalled the works of ancient writers like Herodotus (around 490–425 B.C.), Diodorus Siculus, and Strabo, all of whom had described a mysterious Apis cult in ancient Egypt. In the first chapter of his *Geography*, Strabo (63 B.C.–A.D. 26) wrote that[2]

> Memphis itself, the royal residence of the Aegyptians, is also near Babylon; for the distance to it from the Delta is only three schoeni. It contains temples, one of which is that of Apis, who is the same as Osiris; it is here that the bull Apis

is kept in a kind of sanctuary, being regarded, as I have said, as god. . . .

There is also a Serapium at Memphis, in a place so very sandy that dunes of sand are heaped up by the winds; and by these some of the sphinxes which I saw were buried even to the head and others were only half-visible.

Can you believe it? Partially buried sphinxes, Memphis, the Apis bull, and the Serapis temple. Mariette was in the right place! In the forty-volume *Historical Library* authored by Diodorus Siculus, who lived in the first century B.C., Mariette had read about the sacred bull called Apis and its lavish burial.[3]

A lavish burial? So far, no one had found burial sites for bulls in Egypt. Auguste Mariette forgot all about the mission with which he had been entrusted by his French colleagues; he forgot about the Coptic patriarchate and the papyri which he was supposed to copy. He had his own mission. Without hesitation, he hired thirty workers and ordered them to shovel away the sand from the small dunes that were protruding from the desert at short intervals. Auguste Mariette uncovered a large number of sphinxes, one of them about every six meters; an entire parade of 134 statues was unearthed. Strabo had been right after all!

Among the ruins of a small temple, Mariette found some stone tablets with pictures and inscriptions. They depicted Pharaoh Nectanebo II (360–342 B.C.) dedicating the temple to the god Apis. Mariette was certain that the tombs of the lavishly buried Apis bulls had to be nearby.

The following weeks were occupied by a feverish search. Mariette made a wealth of discoveries. From the desert sands he retrieved likenesses of falcons, gods, and panthers. In a chapel of some sort, he unearthed a limestone statue of an Apis bull. The bull sculpture triggered some unexpected reactions among the women of the neighboring villages. During a lunch break, Mariette observed fifteen girls and women taking turns on top of the bull. They positioned themselves on the bull's back and began moving their stomachs and thighs rhythmically against the statue. These exercises, Mariette was told, were a panacea against infertility.

In his search for the entrance to the bull tombs, Mariette discovered hundreds of small figurines and amulets. Rumors started to circulate in Cairo that the nervous French archaeologist was stealing golden statues. Egyptian government soldiers on camels rode into camp to bar Mariette from further excavations.

Mariette swore, cursed—and negotiated. His bosses in Paris, delighted by the reports and treasures they received from Mariette, sent him another thirty thousand francs and intervened diplomatically on his behalf with the Egyptian government. On June 30, 1851, Mariette was granted permission to continue his work. Impatiently, he resorted to using dynamite so that he could listen to the sound waves reflected off the ground during the explosion.

Where Are the Mummified Bulls?

On November 12, 1851, a large rock gave way under the weight of Mariette's feet. As if on an escalator, he slowly slid down into an underground vault. After the dust had settled and some torches had been lowered into the chamber, Mariette found himself in front of a recess in the wall which contained a gigantic sarcophagus. He did not doubt for a second that he had accomplished his goal. The sarcophagus was certain to harbor one of the sacred Apis bulls. As he moved closer to light the niche with his torch, he saw the huge cover stone of the sarcophagus. It had been forced out of its original position. The sarcophagus was empty.

In the weeks that followed, Mariette examined the mysterious vaults methodically. The main vault was close to three hundred meters long, about eight meters high, and three meters wide. Large chambers branched off on both sides. Each contained a granite sarcophagus, solidly anchored at the base. A second vault, as large as the first, was opened. The twelve sarcophaguses found here had the same oversized dimensions as the twelve discovered in the first vault. These are the dimensions of a sarcophagus: length = 3.79 meters; width = 2.3 meters; height = 2.4 meters (without the cover); width of the sarcophagus wall = 0.42 meters. Mariette estimated the weight of each sarcophagus at roughly seventy tons, with an additional twenty or twenty-five tons for the cover stone. Humongous! All the cover stones had either been partially moved aside or pushed from the sarcophagus altogether. Not a trace anywhere of the lavishly buried sacred bull mummies.

Mariette assumed that grave robbers or monks from the nearby St. Jeremiah monastery had beaten him to the sepulchers. Bitter and infuriated, he continued digging. More vaults were forced open. They contained wooden sarcophaguses from the nineteenth dynasty (1307–1196 B.C.). When a large boulder threatened to halt his progress, Mariette once more resorted to dynamite. The explosive ripped a huge hole in the ground, and in the light of their torches, the men beheld a giant wooden sarcophagus below. The explosion had shattered its cover. Once the debris was removed, Mariette encountered the remains of a mummified man. According to Mariette, the mummy's face was covered by a gold mask; around his neck he wore a gold chain with a small column made of green feldspar and red jasper. Another chain featured two jasper amulets inscribed with the name of Prince Khaemwese, who was one of the sons of Ramses II. Eighteen statues with human heads were scattered nearby, all with the inscription "Osiris-Apis, Great God, Lord of Eternity."[4]

But it was not until the 1930s that this mummy, which Mariette assumed to be the remains of a prince, was scientifically examined. When British Egyptologists Sir Robert Mond and Dr. Oliver Myers cut apart the bandages, they were engulfed by the nauseating smell of bitumen (asphalt). The formless mass was riddled with tiny bone fragments.

Where were the sacred bulls? During the summer of 1852, Mariette discovered additional Apis sarcophaguses in a new vault. The oldest among them were dated around 1500 B.C. None of them contained mummified bulls!

Finally—it was on September 5, 1852—Mariette encountered two sarcophaguses that had not been tampered with. The dust on the ground revealed the footprints of the priests who had carried the sacred bulls to their tombs thirty-five hundred years earlier. A gold-plated statue of Osiris was standing guard over the niche.

The floor was covered with gold tiles which had peeled off the ceiling over the course of time. On the ceiling, Mariette recognized depictions of Ramses II (1290–1224 B.C.) and his son offering a sacred potion to the god Apis-Osiris (presented in this case as an amalgam of both). With great difficulty, the lid was hoisted off the sarcophagus with crowbars and winches. Auguste Mariette wanted to make certain that he would indeed find an Apis mummy and therefore used extra caution. His first concern was for the head of the bull, but he found none. The sarcophagus contained a foul-smelling bituminous mass which disintegrated under the slightest pressure. This mass was permeated with a large number of tiny bones, which apparently had been fragmented at the time of burial. Amidst the array of random bone fragments, Mariette was fortunate to find fifteen small figurines.[5]

The same disenchanting experience repeated itself with the second sarcophagus. No skulls of bulls, no large bones; on the contrary, he found even more tiny bone fragments.

The vaults beneath the sands of Saqqâra never yielded any sacred Apis bulls, contrary to everything the tourists are told and contrary to most of the pertinent scientific literature. Today, the vaults are called the Serapeum, a Greek synthesis of the words Osir and Apis-Serapis.

Auguste Mariette, the restless searcher, who fought many battles with the Egyptian authorities, returned to Egypt after a brief trip to Paris. He had grown unaccustomed to the stuffy museum air at home. In 1858, the Egyptian government appointed him chief supervisor of all historical excavations in Egypt, based on the recommendation of Ferdinand de Lesseps, the builder of the Suez Canal. The energetic Frenchman threw himself into his work. Under his supervision, excavations were going on at forty different locations simultaneously, involving up to 2,700 workers. Mariette was the first Egyptologist to create a detailed catalog of all his findings. He founded the world-renowned Egyptian Antiquities Museum and in 1879 was awarded the title of pasha. The libretto to the opera *Aïda,* which Guiseppe Verdi composed for the opening of the Suez Canal, honors Auguste Mariette. Unknowingly, thousands of tourists pass his grave every day. His sarcophagus is located in front of the Egyptian Antiquities Museum in Cairo.

Sarcophaguses with Fake Mummies

In the eyes of conservative archaeologists, there is little doubt that the giant sarcophaguses at the Serapeum at one time housed the mummies of sacred bulls. An expert recently hissed at me, "What else should have been in them? Nuclear waste?" I doubt that, dear gentlemen, but the solution to the mystery might come from totally unexpected corners. In order to track down the perpetrator with ironclad detective work, I first have to present some curious facts.

In addition to the sacred Apis, the Egyptians worshiped two other, less well-known bulls named Mnevis and Buchis. Strabo mentions the city of Heliopolis, "which is situated upon a noteworthy mound; it contains the temple of Helios, and the ox Mneuis, which is kept in a kind of sanctuary and is regarded among the inhabitants as god, as is Apis in Memphis."[6]

Mnevis was a bull without markings but with black tufts of hair that were brushed the wrong way. A letter written by one of the temple priests at Heliopolis describes the actual mummification of the Mnevis bull. The priest was confirming the receipt of twenty arm lengths of fine linen to wrap Mnevis. Excavations at Heliopolis, the city of the sun god Atum-Re, indeed revealed sepulchers of Mnevis bulls: All of them had been destroyed, robbed, and pillaged. Up to this day, not one of the Mnevis tombs has been found intact.

The Buchis bull was worshiped in central Egypt, not far from the modern-day city of Luxor. As is often the case in archaeology, the Buchis catacombs were discovered by accident. The British archaeologist Sir Robert Mond had heard that a bronze statue of a bull had been uncovered near the small town of Armant. This small town is identical to the former temple city of Hermonthis, which in ancient Egypt was also called the southern On (as opposed to the northern On, namely Heliopolis). Sir Robert Mond calculated that there may well have been a bull-worshiping cult in the southern On, just as there had been in its northern counterpart. The bronze statue confirmed his suspicions. He launched a methodical search.

Like Auguste Mariette with his discoveries at Saqqâra, the British team unearthed huge sarcophaguses in subterranean vaults located beneath the dilapidated temple ruins of Hermonthis. The sarcophaguses at Hermonthis were similar to those at the Serapeum in that they were built into niches on both sides of the main entrance. In consideration of the fact that the whole structure, containing thirty-five niches, was dedicated to the Buchis bulls, it was named "Bucheum."[7] Not far from it, Sir Robert discovered a second set of vaults, which was named Baqaria. Both structures were in extremely poor condition. Once more, grave robbers had beaten the archaeologists to the loot; moreover, some of the chambers were partially submerged in water and the mummies presumably had been eaten up by millions of white ants. The bronze statues that were found were totally corroded; artifacts made of iron had been atomized. Sir Robert Mond reported that the most well-preserved body they found, at the very end, was that of Baqaria 32. The mummy was handled with extreme caution and every detail was recorded. The position of the mummy did not indicate a resting bull but rather a jackal or a dog. No bones had been broken.[8]

All of this sounds strange and confusing. The existence of the sarcophaguses is the only fact that can be ascertained. They were found at the Serapeum, in vaults below Heliopolis, at the Bucheum, in Baqaria, and also in Abu Sir near Giza. The sarcophaguses were either empty or contained a foul-smelling bituminous mass filled with tiny bone fragments.

Even more confusing is the fact that a human mummy with a gold mask was found instead of the anticipated sacred bulls. Much later we learn that the bandages did not cover a human corpse either, but once more concealed a malodorous asphalt mass. And finally, the mummies which supposedly were bulls in reality turn out to be jackals or dogs.

Unanswered questions abound. British archaeologists Mond and Myers had some of their artifacts from the Bucheum and from Baqaria chemically analyzed. One piece of white glass contained 26.6 percent aluminum oxide, a value far too high for ordinary glass. An artificial eye made of clay contained an excessive amount of calcium, and the white of the eye, which had been assumed to be faience, turned out to be neither Egyptian faience nor glass. (In contrast to genuine faience, Egyptian faience consisted of a fine quartz sand with a glass glaze. The Egyptians used it to make jewelry, most notably beads.)

The (coverless) bull sarcophaguses were made from a single block of Aswan granite. Aswan is located approximately six hundred miles from the Serapeum. The mining, polishing, and transportation of even a single sarcophagus, weighing ninety to one hundred tons including the lid, would have been a superhuman feat. Moreover, these monstrously heavy and bulky coffins had to be pushed, pulled, or rolled to the prepared vault and anchored into the wall recesses. Such accomplishments in terms of organization and technology underline the degree of importance which the Egyptians conferred upon the contents of the sarcophagus. But lo and behold: The priests would cut up and virtually pulverize the recently mummified bulls into tiny bone fragments, mix them up with a large, sticky blob of bitumen (asphalt), stir in some religious statues and amulets, hoist the smelly concoction into a splendid sarcophagus, and put on the lid. Just like that!

If this had been the scenario, the ancient Egyptians certainly could have saved themselves the trouble of making and transporting the gigantic sarcophaguses. There was no need for colossal granite coffins to preserve some bone fragments—without heads or horns—for several centuries. Furthermore, the experts agree that the ancient Egyptian priests would have never dissected one of the sacred bulls—it would have been blasphemy. Sir Robert Mond indicates that it was inconceivable in ancient Egypt for a mummy to be buried in any form that would not have left its physical body intact.

Nevertheless, it appears that this happened time and time again. The underground structures near Abu Sir yielded two splendidly embalmed bulls. The linen bandages crisscrossing the entire body were tied down with strings and were unsevered. Hopes were high that finally two well-preserved mummies of sacred bulls had been found; the bandages, after all, were topped by a head with horns on it. Very carefully, the French specialists Lortet and Gaillard cut through the ancient strings and gradually peeled away the numerous layers of bandages. They were flabbergasted. The inside of the mummy exposed a hodgepodge of various animal bones from different species. The second mummy, two and a half meters long and one meter wide, had all the outward appearances of a bull, but it contained the bones of at least seven different animals, some of them calves, some bulls.

All the vaults designated for the bulls had been tampered with. Had they been plundered by grave robbers? Or had monks smashed the contents of the sarcophaguses into tiny pieces? Traditionally, grave robbers have been interested, almost exclusively, in money and precious stones; they never cared about mummified steer. Besides, there is no reason to suspect that grave robbers would have placed an assortment of animal bones into some kind of pseudo-mummy. It would appear a bit more conceivable that God-fearing monks carried their missionary zeal a little too far, provided they knew all the entrances to the sacred bull cemeteries. The monks, in their religious fervor, could have hoisted the lids off the sarcophaguses and proceeded to crush the contents with heavy iron bars, sort of like crushing grapes. Unfortunately, this explanation does not hold water either. Such a Christian rampage would have left traces—shredded bandages, for example. Icons of the Egyptian gods would have been smashed or melted. Most likely, the pious monks would have thrown a Christian cross into each of the sarcophaguses to exorcise the heathen demons, or they would have left some religious statues of their own in the underground galleries. There is no indication of any such interference. Where are the mummies of the sacred Apis bulls?

Contradictory Sources

According to Herodotus (around 490–425 B.C.), who traveled extensively in Egypt around the year 450 B.C., talking to the local priests, our search for Apis mummies is bound to be in vain. Herodotus reports that the Egyptians simply ate their sacred bulls.

> Male kine are reckoned to belong to Epaphus, and are therefore tested in the following manner:—One of the priests appointed for the purposes searches to see if there is a single black hair on the whole body, since in that case the beast is unclean. He examines him all over, standing on his legs, and again laid upon his back; after which he takes the tongue out of his mouth, to see if it be clean in respect of the prescribed marks (what they are I will mention elsewhere); he also inspects the hairs of the tail, to observe if they grow naturally. . . .

The following is their manner of sacrifice:—They lead the victim, marked with their signet, to the altar where they are about to offer it, and setting the wood alight, pour a libation of wine upon the altar in front of the victim, and at the same time invoke the god. Then they slay the animal, and cutting off his head, proceed to flay the body. Next they take the head, and heaping imprecations on it, if there is a market-place and a body of Greek traders in the city, they carry it there and sell it instantly; if, however, there are no Greeks among them, they throw the head into the river. . . .

The disembowelling and burning are, however, different in different sacrifices. . . . When they have flayed their steer they pray, and when their prayer is ended, they take the paunch of the animal out entire, leaving the intestines and the fat inside the body. They then cut off the legs, the ends of the loins, the shoulders, and the neck; and having done so, they fill the body of the steer with clean bread, honey, raisins, figs, frankincense, myrrh, and other aromatics. Thus filled, they burn the body, pouring over it great quantities of oil. Before offering the sacrifice they fast, and while the bodies of the victims are being consumed they beat themselves. Afterwards, when they have concluded this part of the ceremony, they have the other parts of the victim served up to them for a repast.

The male kine, therefore, if clean, and the male calves, are used for sacrifice by the Egyptians universally; but the females they are not allowed to sacrifice, since they are sacred to Isis.[9]

This according to Herodotus. If he were correct, our examination of the bull necropolises would be pointless. Why would these people have gone through all the trouble with the granite sarcophaguses if the animals were intended to end up on the priests' plates? Paradoxically, the same writer, Herodotus, describes in a different passage how a bull was embalmed and how its entrails were separated from the body by flushing cedar oil through its intestines. The ancient writers were generally rather contradictory in describing the sacred bulls. While the bulls are consumed by the priests in the works of Herodotus, they are lavishly buried in the writings of Diodorus Siculus. On the other hand, ancient Roman authors like Plinius, Papinius Statius, and Ammianus Marcellinus agree that the bulls were drowned in a sacred well.

Drowned—eaten—embalmed—cut up: Which do we prefer?

One ancient Egyptian source, the "Papyrus Apis," details the procedure for mummifying a sacred bull. It describes the process step by step and notes how many priests were supposed to be standing in which positions; it also prescribes the way in which the bandages were supposed to be wrapped—left and right, from the top and from the bottom, as well as across. After being cleaned with water and oil, the bull had to be covered with natron until it was completely dried. During the entire ceremony, one of the priests was required to stand in front of the bull, chanting and reciting prayers and supervising the embalming process to ensure that every movement followed the prescribed routine. Once the animal was wrapped in a few thousand yards of linen, the skull was covered with plaster. A gold disk was placed between the horns, symbolizing the bull's relationship to the sun god. Finally, glass eyes were jammed into the eye sockets and the finished mummy was carried to its prepared place in the vault in a solemn procession. All of this has been described in detail. What went wrong?

Who Was Omar Khayyám?

An acquaintance of mine had invited me to dinner in a typical Egyptian restaurant. We had rice, chicken, steamed brown beans mixed with onions, and the favorite Egyptian vegetable, Molokhiyya. Its leaves are tasty and juicy, and it is mainly used in spicy soups or vegetable casseroles. My companion and I were enjoying a heavy, flavorful wine as he told me about the terrible times under Caliph El Hakim, who ruled over Cairo from 996 until 1021. At the caliph's command, every person found eating Molokhiyya was executed on the spot. Sadistically, the caliph did not merely aim to reeducate his subjects, he actually enjoyed their suffering. Ever since the rule of Caliph El Hakim, it has been impossible, politically, for any Egyptian government to curtail the amount of Molokhiyya grown.

While my companion happily consumed large quantities of the green leaves, my eyes caught sight of the label on the wine bottle, which read "Omar Khayyám." Who was Omar Khayyám?

"I believe it must be the name of the wine grower or the wholesaler," my companion offered.

One of the waiters, who had overheard our conversation, disputed this claim immediately. "Omar Khayyám was a former ruler of Egypt," he said.

Suddenly, the headwaiter appeared at the table and dismissed his subordinate with a brusque gesture. "Omar Khayyám was a famous general," he insisted. Now a guest at the next table got involved as well. "Omar Khayyám? He was a bedouin chieftain, wasn't he?" I was beginning to wish I had never asked. The entire restaurant got into the act, wildly speculating about Omar Khayyám. Soon the atmosphere resembled the chaos of a stock exchange floor. "An admiral," someone shouted. "The founder of the zoological garden," another chimed in. "What are you talking about?" interrupted an older gentleman with a wide gap in his teeth. "Omar Khayyám engineered the Aswan dam. . . ."

Many days later, during a conversation with Dr. Holeil Ghaly, the man in charge of excavations at Saqqâra, I asked him, half in jest: "Who was Omar Khayyám?"

The district chief of archaeologists smiled and grabbed an encyclopedia. "Omar Khayyám," he read. "Persian poet, mathematician, astrologer, 1048–1122, dealt with philosophical questions, wrote flowery love poems."

One just has to find the right people to ask.

A Pyramid Is Found

I was talking to the right person. Dr. Holeil Ghaly is not just any Egyptologist. According to his title, he is "director of antiquities at Saqqâra." He is a smart, pleasant, and knowledgeable man who speaks a number of languages and who even admitted to having read some of my books.

"Imagination is important," he said, "even in archaeology." I wish there were more archaeologists like him!

Saqqâra has the most extensive archaeological excavation site in Egypt, the largest excavation area in the world. It starts at the border of Giza, near Abu Sir, and extends almost forty miles south along the Nile. During the winter months, the desert sand is dredged by several international teams trying to wrest hidden secrets from the rocky ground below the sands. As late as spring of 1988, a French team from the Collège de France discovered two previously unknown pyramids from the era of Pepi I (2289–2255 B.C.).

"Would you like to see the pyramids?" Dr. Ghaly asked. We were skimming over the sand dunes in his Jeep, passing the portion of Saqqâra that is open to tourists. During the ride, I learned that Pharaoh Pepi had been known for years. He succeeded Titi (2323–2291 B.C.), who was the founder of the sixth dynasty. Titi and Pepi—I wish our current politicians had simple names like these! The pyramid of Pepi I is situated in the southern part of Saqqâra, and the two pyramids found by the French team are not far from it. One is a pyramid reserved for Pepi's family. What is there to discover? I thought. Don't pyramids naturally protrude from the desert sand?

It was almost four o'clock in the afternoon. The heat covered us like a sheet of hot embers, searing the pores of our skin and our scalps. The Jeep came to an abrupt halt in front of a large hole in the ground. No trace of a pyramid anywhere. Dr. Ghaly, my assistant Willi Dünnenberger, and I approached the edge of the hole. I was speechless—not from the heat, which was heavy enough to cut with a knife, but from the sight which opened itself to my incredulous eyes about ten meters below. We have grown accustomed to the sight of pyramids while standing in front of them, admiring their perfect contours against the backdrop of a clear blue sky. This was different. Like time travelers from a distant dimension, we were standing ten meters *above* the ruins of the pyramid, which evidently had been used as a convenient quarry by nearby residents for centuries. Nevertheless, two sections of the pyramid had remained intact, displaying perfectly polished and jointed stone blocks.

"How long have you been digging here?"

"The French team has been working on it for the past six months, together with Egyptian archaeologists and a total of a hundred and eighty workers," explained Dr. Ghaly. "Now, during the summer, it is impossible to excavate because of the heat."

The archaeologists from the Collège de France had located the pyramid under thick layers of sand and rock with the help of electronic devices. A number of new methods are available which the likes of the nineteenth-century German archaeologist Heinrich Schliemann could never have imagined. A magnetometer is used to determine the magnetic field of a given site. The unit of measurement is called a "gamma." The earth's magnetic field varies between twenty-five thousand gamma at the equator and seventy thousand gamma at the poles. With sophisticated processes, the gamma value of a particular site is determined and compared with actual readings taken at the site to verify if the value is consistent. If irregularities occur because of metals or hollow spaces below the surface, a ground-penetrating radar (GPR) is added. It is similar to sonar. A transmitter sends high-frequency impulses into the ground, which are reflected and measured by a special antenna. A portable computer monitors the impulses and translates them into waves and lines on the screen. If something unusual has been traced under the soil, the location of the foreign object can gradually be narrowed down with the help of GPR. In this manner, the French archaeologists found something without ever using a spade. Similarly, a team of physicians and archaeologists from the University of California at Berkeley has been working on a detailed map of the underground structures in the Valley of the Kings for the past ten years.[10]

Tombs whose existence had fallen into oblivion for thousands of years are being located now; underground vaults are detected through radar. It is quite conceivable that we will be able to locate more archaeological treasures in the next ten years than have been found in the past one hundred years combined. If modern-day treasure hunters search in the right places and do not spare any effort or expense, they are bound to find something. Unfortunately, some religious and political groups object to these archaeological ventures. It is generally those who are behind their times who are afraid of what the past might uncover.

On the way back, I inquired after the meaning of the word "Saqqâra."

"The word is used in the old Egyptian language. Saqqâra comes from the word 'jackal,' " Dr. Ghaly explained.

"How old are the oldest of the artifacts found in Saqqâra?"

Dr. Ghaly, a very youthful-looking man, tilted his head. "The history of Saqqâra begins with the first dynasty around 2920 B.C. and extends past the birth of Jesus Christ. There are even some prehistoric findings."

I was still wrapped up in the mystery of the Apis bulls. So I said to Dr. Ghaly, "I have studied Auguste Mariette's excavation report very thoroughly. Are you aware that Mariette never found any bulls in the Serapeum?"

Dr. Ghaly thought for a moment. "Yes, I am aware of that."

"Can we expect any great surprises to result from the work being done at Saqqâra?"

The Egyptologist smiled benevolently and flashed his shiny white teeth, which sparkled like ivory under his black hair. "We assume that we know about twenty percent of Saqqâra. Eighty percent remains untouched under the ground."

Good grief! I thought. Twenty percent and so many unanswered questions. What kind of surprises may lie in store for us in the future? How many visitors to the pyramid of Djoser, the pyramid district of the Pharaoh Unis (2356–2323 B.C.), or the splendid tomb of the nobleman Ti in Saqqâra are aware that the ground below their feet conceals a maze of thousands of subterranean tunnels? How many of these curious travelers, exhausted by the heat, slurping sweet tea or lukewarm Coke in the tourist tent, are told that millions (!) of mummified animals of various species are entombed below the sands of Saqqâra? And who would be there to tell them? In essence, they are sitting on top of an oversized version of Noah's ark.

In my opinion, the monumental sarcophaguses for pseudo-animals have a central purpose. But patience, please! I am homing in on these monsters for whom the ancient Egyptians spared no effort! Where does this obsession with mummification come from? The mummification of humans may be somewhat understandable. But animals?

Body, Ka, and Ba

Much is known about the religious ideas of the ancient Egyptians, from pyramid texts, an abundance of tomb inscriptions, papyri, and books by ancient authors like Herodotus. When the god Khnum (who is depicted with a ram's head) created man, he made two parts: body and ka. The body is mortal, the ka is immortal. The ka is part of the all-encompassing, universal spirit, the vibrations, so to speak, that generate all life. The body is nothing but matter; it would be lifeless without the ka. In contrast, the ka is spiritual, omnipresent, and eternal. However, the ka does not correspond to our notion of a soul. Reinhard Grieshammer, a leading authority on the subject, wrote that people wanted to perceive the ka as a person's double or some kind of guardian angel. The only thing that is certain is that it embodies some kind of force or power. Grieshammer explains that, according to texts and drawings, the aspect of man which we call the ka enters the body at birth.[11]

In addition to the ka, each person also has a ba. This identifies a condition which is not present until body and ka are united. The ba could be likened to one's conscious awareness, individual conscience, psyche, or store of knowledge. Once the body dies, the ka is united with the ba. The ancient Egyptians used to say about a deceased that "he went to his ka." The body is nothing but an empty shell, but the ka and the ba are united, are eternally bonded, and appear before the gods and their ancestors in a different dimension.

This ancient theory, which has been promulgated by

the religions of this world for centuries in one way or another, has been rejuvenated lately. The names have changed, but the idea is the same. Physics interprets matter ultimately as vibrations. The world of the atom and the subatomic particle, which are at the root of everything, is the dimension of radiation and vibration. Example: The electron, a component of every atom, pulsates 10^{23} times per second. That's a ten followed by twenty-three zeros. Physicists, ever searching for the universal formula which could explain everything and combine all knowledge, have no idea where these vibrations originate, or what fuels their motion. On the other hand, esoterics and philosophers, people bound by the limitations of reason and emotion, proclaim that all is one, that everything is somehow interconnected.

Tree, beast, and human all have vibrations—ka—but the plant and the beast are released from personal liability. The actions of a tree, for instance, are neither right nor wrong, neither good nor bad, neither logical nor illogical. As a result, it never develops a psyche, an individual responsibility. It lacks the ba. Only the trinity of body, ka, and ba gives each person his or her unique personality and distinguishes him or her from all others. No two people, not even identical twins, assimilate, tolerate, or register the same experience the same way; no two people suffer or rejoice with the same intensity. We are all humans, made from the same genetic material—but each of us is unique. We have *become* what we are.

So far, so good. But none of this explains why anyone would want to mummify a dead body, an empty shell without ba or ka. Among ancient Egyptians, the peculiar notion arose that the ka was still joined to the body even after death, that the ka needed the body to return. If the ka and the ba were to fare well in the afterlife, the body had to be sustained. It is unknown how the Egyptians and other cultures practicing mummification arrived at this peculiar conclusion, because it ultimately contradicted their own beliefs. According to their religion, the body was worthless ballast once the ka and the ba left it. This new idea that the body should be preserved was directly responsible for the custom of mummifying bodies and for the heavily fortified burial vaults. The vaults were designed with traps and convoluted paths to protect them against enemies and robbers. The amount of treasure buried with a deceased person depended on his or her wealth. Not only gold, precious stones, and nonperishable foods were added to the lightless tomb, but also the deceased's favorite objects, toys, and jewelry, and even his or her bed and tools. It was considered essential that the deceased feel at home and carry with them enough objects of value for the countless offerings required on their long journey through the other world.

All of this is true and documented by the artifacts found in the tombs—but it is also wrong and illogical. I am tempted to ask: How dumb do we think the old Egyptians were? Or else: What have *we* missed in our interpretation of burial sites and ancient texts? All explanations concerning the pomp and circumstance which the ancient Egyptians afforded their dead are built on shaky ground, contradicting all practical experience and interpretation. Why?

Through the ages, graves, even the heavily protected sepulchers of pharaohs, have been plundered by greedy descendants. This is nothing new; it occurred way back in the heyday of these extensive, eccentric, labyrinthian burial vaults. By the beginning of the eighteenth dynasty (around 1500 B.C.), the tombs of virtually all political rulers had been plundered. We know from inscriptions that the pharaoh Rhombi (1319–1307 B.C.) ordered the restoration of the tomb of his predecessor, Thutmose IV (1401–1391). Thutmose had only been in the grave for eighty years. The pharaohs and priests knew, without a doubt, that he had neither taken his treasures and favorite belongings into another dimension nor used them as offerings on his way through the other world. Instead of drawing the logical conclusion that this whole mummy business was a sham, particularly since it went counter to the religious concept of the spiritual and immortal ka, the priests doubled their efforts. They moved their burials to the Valley of the Kings near Thebes, where they chiseled underground vaults into the mountains, securing them with traps and monstrous boulders. And the dead were pampered even more. The tomb of Tutankhamen (1333–1323 B.C.), which by some fortuitous circumstance avoided being pillaged, is a clear indication of this.

It just doesn't add up!

The Sleeping Dead

Twenty-six years ago, in my book *Chariots of the Gods?*, I expressed my suspicion—far-fetched at the time—that the ancient Egyptians chose their burial method with not a spiritual but rather a physical reincarnation in mind.

"That is why the provisioning of the embalmed corpses in the burial chambers took such a practical form and was intended for a life on this side of the grave. Otherwise what were they supposed to have done with money, jewelry, and their favorite articles? And as they were even provided in the tomb with some of their servants, who were unquestionably buried alive, the point of all the preparations was obviously the continuation of the old life in a new life. The tombs were tremendously durable and solid, almost atom-bomb proof; they could survive the ravages of all the ages. The valuables left in them, gold and precious stones, were virtually indestructible."[12]

At the time, I referred to a book by the physicist and astronomer Robert C. W. Ettinger, who outlined a method to preserve corpses for a later revival.[13] And today?

In the United States—where else?—an organization has been formed which calls itself the American Cryonics Society (ACS). Founder and president of this organization is a mathematician named A. Qaife, who steadfastly refuses to perceive death as an inescapable reality. It is the purpose of his society to preserve and freeze corpses so that they can be thawed out later—after decades, centuries, or even millennia. These kinds of experiments are in an advanced stage as far as animals are concerned. ACS member Dr. Paul Eduard Segall confirms having frozen his own dog and revived it after fifteen minutes, at which time the dog was perfectly fine and wagging its tail! The same process has been performed hundreds of times with hamsters. Twenty percent of the animals survived the deep freeze. Cats, fish, and turtles have also been frozen with some success. The animals are drained of their blood, which is then replaced with an antifreeze solution. (Blood would freeze and the cells would tear.) The bloodless bodies are then stored in liquid nitrogen in special tanks at minus-196 degrees Celsius. With human beings, it is conceivable that the brain and other sensitive organs would be removed from the body and stored in separate containers. A similar method is used today for the transportation of organs for transplants. Reminiscences of Frankenstein!

Several years ago, I visited a large burial pyramid near Orlando, Florida. Instead of being buried in the ground or burned, the coffin with the remains of the deceased is placed in a refrigerated drawer. Each drawer is labeled with its occupant's personal data. The cause of death is stated as well. The center of the pyramid features a carpeted memorial hall in which the dead person's relatives can visit with their loved one anytime. Elevators connect the numerous floors inside the pyramid. The living and the dead are treated to subdued organ music twenty-four hours a day.

What conclusions would archaeologists draw from these frozen bodies or mummified corpses in hermetically sealed drawers after thirty-five hundred years? That's the time span from which *we* judge the mummifications of ancient Egyptians. Some people object that there is no comparison, because the old Egyptians also endowed their mummies with written blessings and verses for their journey through the netherworld. The Egyptian funeral books are filled with such advice and directions on how to behave in the afterlife. Does this objection hold?

Anybody of sound mind who agrees to be frozen and to have his or her organs and brain preserved in separate containers after death is counting on a *physical* rebirth. This does not prevent relatives from adding pious verses and psalms to the refrigerated box. An inscription might read, "Look forward to life in a better world." Or "In your new life you will be free of the illness that tortured you in this life. May the almighty, eternal God protect you on your journey and bless you."

Future archaeologists would be forced to conclude from such blessings that the deceased believed in a life after death. But is that necessarily so? How can we know for sure what motivated a pharaoh forty-six hundred years ago to have a luxurious tomb built for himself that was to last all eternity? Of course, everybody else wanted to follow the example of the supreme ruler and to be mummified. The original purpose, the hope of a physical resurrection, was quickly forgotten. Promoted by the priesthood, which made a large profit from the business of mummification, a cult began in Egypt which is unequaled in the entire world. The mummification business created totally new professions—embalmer, corpse cleaner, corpse cutter—and a whole industry. Sarcophaguses were made from granite, alabaster, and wood; huge amounts of honey, wax, ointments, oils, and natron needed to be prepared; millions of canopic urns (to preserve internal organs and brains) had to be made; and several million feet of bandages and linens had to be woven.

What has happened to all of these wrapped corpses?

After the empire of the pharaohs was overrun by the Romans, the Egyptian priests abandoned the burial sites. Thousands of tombs were plundered; mummies and wooden sarcophaguses were used for firewood. When

Christian monks appeared in the area in the second century A.D., they destroyed many underground galleries filled with layers of mummies. During the Middle Ages, mummies were the "in" thing in Europe. They were thought to have medicinal powers! Mummy parts, mummy powder, mummy skin, and mummy paste were recommended against paralysis, cardiovascular disease, liver disease, stomach poisoning, epilepsy, and even bone fractures. This newfound market triggered a mass export of mummies from Egypt; European pharmacies squabbled over them. A small amount of "mummy" was an essential part of every first-aid kit at home or on the road; "mummy" was ingested orally or absorbed in the form of ointments or powders. This obsession with mummies as medicine, which lasted for two centuries, was followed by "Egyptomania," a term coined by the French physician and mummy expert Ange-Pierre Leca.[14] Mummies became highly sought after collector's items. They were exhibited in museums and at country fairs and placed like suits of armor in the foyers of aristocratic mansions. Some were unwrapped in public ceremonies. In the nineteenth century, a businessman from Maine began to use mummies as the raw material for his paper-manufacturing operation. He was very upset when the resins and bitumen in the mummies led to a brown discoloration in the paper. Thus was brown wrapping paper born! The brown paper, unsuitable for writing, was distributed to retailers in rolls. The mummy now served as packaging—wrapped for eternity.[15]

Millions of Animals in Bandages

Man is a bundle of fears, joys, sadness, and hope. Parents, lovers, children, friends die. We have no choice but to deal with the reality of death. Do people live on in some form after their death? Are they well taken care of? Are they suffering? Is death the end of it all, or are our actions in this world judged by gods and spirits in the other world? We do not know. Five thousand years of human history have secured no answers to these timeless questions. There is no scientific evidence to support the idea of life after death, of reincarnation. I am well aware of books claiming otherwise. They derive from religious views, philosophy, esoteric theories, or personal accounts. People talk about life on the other side, about unlimited awareness in colorful spheres. People allow themselves to be hypnotized so that they may return to some of their previous existences. I have studied such experiments in detail. I have even volunteered to have myself transported back to previous lives. Some researchers claim to have been able to communicate with the dead through tape recorders; others have managed to conjure up images of dead people on TV screens. In many of these cases, the evidence seems sensible, logical, even convincing. Unfortunately, it is useless to a scientist. Science demands reproducible evidence, unequivocal data that exclude any other interpretation except that there is indeed life after death, that there is a physical rebirth. Personal accounts, with or without hypnosis, are worthless to a scientist.

This persistent search for answers that transcend the boundaries of our lives is due in part to a compulsive internal drive. Our personal history appears to us as too burdensome and painful. And all of this is supposed to be for nothing? A short life followed by a long death? Never! That should not and cannot be! Life must possess meaning beyond death.

The ancient Egyptians were no more immune to such thoughts than we are. And those who seek answers find them. Since we stubbornly refuse to accept the possibility of a final, permanent exit, we cling to a feeble spark of hope arising in our consciousness. There is indeed a way to escape death: reincarnation! Now the unshakable belief in spiritual or physical reincarnation—life in a much better place—becomes the purpose of our lives. Hope gives us wings and carries us past the daily drudgery, pain, aggravation, and injustice. This belief in reincarnation fuels the rise of such modern-day organizations as ACS, just as it fueled religious organizations favoring mummification in the old days.

All of this is understandable; I can empathize with such a thought process. After all, we are dealing here with our own self. But what in the world could motivate a culture to mummify thousands of animals? It has become an almost commonplace occurrence that some wealthy lady has her favorite dog or cat buried like a human being. An abundance of pet cemeteries attests to this. Throughout the ages, our essential loneliness has driven us to seek out a close relationship with a pet. We may call it, literally, puppy love. But is that a reason to mummify hundreds of thousands of crocodiles, snakes, hippos, hedgehogs, rats, frogs, and fish? These animals are not exactly huggable little pets. What follows is a partial listing of animals which the ancient Egyptians mummified:

bull	falcon	hippo
ram	vulture	frog
goat	owl	eel

gazelle	raven	ibis
wolf	cow	eagle
crocodile	sheep	hawk
shrew	antelope	crow
snake	dog	pigeon
cat	baboon	swallow
lynx	weasel	stork
hedgehog	rat	beetle
bat	lion	hoopoe
fish	bear	goose
fish otter	rabbit	scorpion[16]

One of the most famous, and without a doubt most successful, excavators of Saqqâra was Dr. Walter Brian Emery (1903–1971). As a young Egyptologist, he was part of the team which discovered the underground passageways of the Bucheum (with the sarcophaguses for the Buchis bulls) under the temple city of Armant (the southern On). After 1935, he worked almost exclusively in Saqqâra. He located the oldest pharaoh tombs from the first dynasty, including the adjacent sepulchers of the pharaohs' courtiers, who had to give up their lives upon their masters' demise. In the process of excavating a more recent vault from the time of King Ptolemy (330 B.C. until the Roman invasion), Emery uncovered—about one and a quarter meters under the ground—the remains of a young bull which had once been wrapped in burial cloth. About six meters below that, he found a clay jug with a conical lid. Emery cautiously brushed the dirt off the jug and noticed several more jugs just like it on both sides. Some of them bore the sign of the moon god Thoth. All in all, Emery retrieved more than five hundred urns at this site; each one of them contained the mummy of an ibis.

Only a few meters east of tomb 3510 from the third dynasty (2649–2575 B.C.), Emery dug down about ten meters to discover a shaft which was filled, top to bottom, with ibis mummies. The archaeologists could not believe their eyes when they realized that the shaft they had found issued into a long, winding passageway with over fifty tributaries, which in turn diverged into a number of additional shafts. The whole structure turned out to be a maze, several miles long, storing a total of approximately one and a half million ibis mummies![17] Every one of these birds had been carefully prepared and wrapped in bandages before it was put into the vase. The jugs were stacked tightly from the bottom to the very top. The main passageway, with a height of four and a half meters and a width of two and a half meters, would have been easily accessible for a tractor. The same underground maze was described by Paul Lucas, a Frenchman traveling in Egypt at the beginning of the eighteenth century, as being at least four kilometers long. To this day, it has not been explored in its entirety. Today, sand masks the entrances once opened up by Emery. What use do we have for millions of ibis mummies? Of course, a pottery dealer may hit upon the idea of a lifetime any day now. One and a half million vases are looking for a home.

As far as quantity goes, the ibis mummies found at Tuna al-Gebel break all records. Tuna al-Gebel is near the ancient temple city of Hermopolis, about forty kilometers south of Minya. Archaeologists located an underground pet cemetery there covering an area of forty acres. Through two shafts, the excavators entered a veritable city made of rock, featuring streets, dead-end alleys, and intricate chambers crammed with ibis mummies. However, they also found mummies of other animals, such as falcons, flamingos, and baboons. The catacombs contain at least four million ibis mummies. We know that Hermopolis and Tuna al-Gebel, situated seven kilometers to the west, attracted countless pilgrims because of their sacred animal sites even in Greek and Roman times. A stele of Pharaoh Ikhnaton (1365–1347 B.C.) is considered the oldest monument of this necropolis. One thousand three hundred years separate Ikhnaton from the Romans. Just imagine the persuasive power required of a religion for it to have clung to the same icons for such a long time! Theoretically, the origins of the animal necropolis of Tuna al-Gebel may go back even further, another thousand years into the past.

Coffins for Baboons

In Abydos, we know the exact time. Abydos is situated about 560 kilometers upriver from Cairo. The site is of particular interest to archaeologists, because the tombs found in Abydos date back to the first and second dynasty, some five thousand years before our time. Abydos was the central place of worship for the god Osiris, who was in charge of all earthly matters. Osiris is credited with introducing such useful things as farming and winegrowing on earth and therefore was called "the perfect one." Osiris had a brother named Seth, who was envious of his popularity. Seth lured Osiris into a box, cut him up into pieces, and threw the parts into the river Nile. Legend has it that his head is buried in Abydos. The first pharaohs asked to be buried near their venerated god Osiris. Excavations in Abydos yielded royal vaults of superior workmanship, as well as tombs for courtiers, high officials, and even concubines who followed their rulers into the grave. It is unclear whether their eternal devotion was voluntary or not. Whichever, it was a great honor to be buried in Abydos.

Thus, it is very confounding to discover that the sacred soil of Abydos also contains thousands and thousands of dog mummies. At the turn of this century, archaeologists broke into a shaft that had been barricaded with rocks and found a system of subterranean passages, one and a half meters high and two meters wide. The underground paths led to scores of burial vaults stuffed with dog corpses from top to bottom. The dogs had been wrapped in white linen and stacked in rows of ten. It was impossible to move the corpses; the mummies crumbled into dust at the slightest touch. Among the stacks of dog mummies, the archaeologists found some Roman oil lamps dating to the first century B.C. This suggests that dog mummies were buried in Abydos for many centuries, even up to Roman times. Alternatively, it could be submitted that Roman grave robbers left their oil lamps in the musty vaults of Abydos around 100 B.C.

In any case, we encounter mummies of animals wherever we look, and we haven't even started looking very hard. Remember: Only 20 percent of Saqqâra has been explored. The untiring archaeologist Walter Emery, who chanced upon the millions of ibis mummies at Saqqâra, scored another spectacular hit. In the process of clearing a temple dating to the times of Pharaoh Nectanebo I (380–322 B.C.), Emery discovered a small room that provided access to lower-level passageways. Rectangular niches had been broken out of the rock on both sides of the main corridor. Each contained a wooden box with a baboon wrapped in cloth. The animals' feet were enclosed in lime or plaster, presumably to prevent the wooden sarcophaguses from tipping over. In the southeast corner, the main corridor, about two hundred meters long, issued into a long room without any niches. In the light of their lamps, Emery and his crew discovered a steep step descending into a lower-level vault extending endlessly in an east-west direction. One niche after another had been hammered into the rock, and each featured an upright wooden coffin with a mummified baboon. In the process of clearing rubble out of one of the upper galleries, Emery's assistants found plaster prints of human body parts—hands, legs, feet, arms, even wigs and complete heads. The French Egyptologist Jean Philippe Lauer, at the time one of Walter Emery's collaborators and now the preeminent authority on Saqqâra, stated that without a doubt these plaster casts were intended as medicinal votive offerings left by sick pilgrims in hope of a cure. It was their way of either communicating to their god the nature of their affliction and the specific body part involved or expressing their gratitude for having been healed.[18]

Emery, anticipating further surprises, had the baboon galleries systematically cleaned up. He was a restless scientist with the explorer instinct of Auguste Mariette. And sure enough: In one of the lower-level baboon vaults he uncovered a niche which led to another maze of underground passageways. One of the corridors was completely filled with ibis mummies in perfectly preserved ceramic vases. Thousands of these mummies were blocking the entrance.[19] During excavations in 1970–71, Emery found corpses of birds of prey. It was impossible to count the numbers of eagles, falcons, vultures, ravens, and crows. Jean Philippe Lauer, who saw the incredible network of underground vaults with his own eyes, estimates that their numbers went into the millions.[20] As far as we know today, the Egyptians worshiped and mummified thirty-eight different species of birds.

Emery was certain that the subterranean passages were somehow related to the aboveground structures from the third dynasty (2649–2575 B.C.). Fate, however, would not allow him to prove his theory. He suffered a stroke while doing the kind of work that fascinated him his whole life: excavating an archaeological site.

Mummy Rites and Death Magic

If we consider the effort involved in preparing millions and millions of animal mummies, we are bound to take a step back and think: sacred forms of life dedicated to the gods? That must have been it. Hindus, for example, still perceive the cow as a sacred animal. Yet it has never occurred to them to embalm the dead animals with ointments, let them dry, wrap them in elaborate bandages, place them in monstrous sarcophaguses, and bury them in vaults tediously hammered out of a rock. (For the record: The Egyptians also considered the cow a sacred animal. Who borrowed this custom from whom?)

Along the Nile, people did not only mummify birds, baboons, and dogs; the vases also contained ibis eggs, forty or even one hundred at a time, each individually and carefully wrapped in cloth. More than two hundred thousand mummified crocodiles were buried in the Tebtynis necropolis, an underground cemetery west of the Nile near the oasis Fayyum. Archaeologists found clay urns containing meticulously wrapped crocodile eggs scattered among the decomposed crocodile corpses, partly eaten by insects. Ancient texts (such as those of Herodotus and others) describe a gigantic maze of sacred crocodile remains: the Sucheion. To this day, the location of the Sucheion has remained a mystery.

The old Egyptians did not even spare snakes and frogs from mummification. Various types of venomous snakes, which were abundant in Egypt at the time, were anointed with sweet-smelling potions, wrapped in narrow linen bandages, and placed in long wooden sarcophaguses. Mummified frogs, all wrapped up in cloth, were jammed into small bronze containers. Not to be left out, the priests of Esna, a town about fifty kilometers from modern-day Luxor, specialized in the mummification of fish. Thousands of fish, from the smallest to the biggest, all carefully wrapped, were found in a fish necropolis ten kilometers west of Esna.

From a modern point of view, this whole mummy business can only be interpreted as a religiously motivated ritual on the part of the Egyptians. They venerated their animals and believed that even the poor beasts had a ka, and that they would need their earthly bodies in the afterlife. Economically, the whole custom was foolish. Sarcophaguses and vaults were filled with tremendous amounts of valuables and precious metals; the elaborate mummification of all these animals required an incredible amount of labor. What for? To preserve dried corpses?

The Egyptians knew, with the benefit of a thousand years of experience and from their everyday observation, that these mummies were not going anywhere. None of the empty, bandaged shells magically returned to life, no dead crocodiles tunneled themselves out of their burial cloths, and never did the barking of resurrected dogs interrupt the dreary silence of a necropolis. There is absolutely no doubt that the Egyptians practiced their animal worship in prehistoric times; it was not a custom introduced by the pharaohs' priests. What belief or misbelief was so powerful that it was able to endure in Egypt for thousands of years?

Even ancient authors pondered the question. Diodorus Siculus writes:

> Everything the Egyptians do in their veneration of the sacred animals is strange and unbelievable, and this presents great difficulties to anyone investigating its causes. Their priests hold certain secret doctrines concerning these matters, which we explained earlier in our discussion of their gods. But most of the Egyptians allege one of the following three explanations, the first of which is wholly fabulous and quite in keeping with old-fashioned simplicity of beliefs: for they say that the first gods who existed were few in number and oppressed by the multitudes and by the lawlessness of earthborn men; but that they escaped this savagery and cruelty by adopting the forms of various animals. Later, however, when they had come to be masters of the universe, they showed their gratitude to the agents of their earlier deliverance by sanctifying those species of animals which they had imitated, teaching men how to care luxuriously for the living and perform funeral rites for the dead. The second reason they offer is that the Egyptians of old, through lack of discipline in their army, were defeated in many bat-

> tles by the neighboring peoples, until they conceived the idea of carrying ensigns before the divisions of their host. Therefore, they continue, the leaders fashioned images of the animals which they have now come to worship, and bore them elevated upon javelins; and in this way every soldier was able to recognize his own unit in the line of battle. . . .
>
> The third explanation they put forth concerning this question is the service that each of the animals provides for the benefit of mankind and the life of the community.

Diodorus Siculus emphasizes repeatedly that this is the interpretation given by the people. The priests' knowledge of the origins of animal worship had to be "kept secret." Even back then!

Lucian (around A.D. 120), a Greek author who, in his old age, became imperial secretary in Egypt, writes that the Egyptian animal cult is based on astrology. He maintains that the people in different districts of Egypt worshiped different signs in the sky and that they transferred these signs onto their local animals. However, other ancient writers dispute this claim. They assert that the animals were venerated out of fear and terror or because they performed miracles. Diodorus Siculus reports one such instance, the story of a king named Menes, who was chased into Lake Moeris by his own dogs and was saved by a crocodile, which carried him to the other side of the lake.

We are tempted to dismiss such stories as legends and fairy tales born of overactive imaginations. But could they contain a grain of truth? Could it be some misinterpreted basic truth whose real meaning is known only to priests and insiders? Dr. Theodor Hopfner, an expert who studied the animal cults of ancient Egypt seventy years ago and who was familiar with all the relevant old texts, concluded that none of the available facts would explain why the Egyptians should have assumed that their gods would be incarnated in animals. Neither did Hopfner believe that the animal cults were based on the supposition that the souls of deceased humans would be incarnated in animals. He saw no signs to indicate that the Egyptians believed in a transmigration of the soul.[21]

Now what? Here is an interesting fact: Only certain animals within a given species were considered sacred. The priests did not place the holy seal on just any gazelle, dog, cow, or bull; they only chose those with unmistakable characteristics. Herodotus describes the black-and-white Apis bull as having the following marks: "He is black, with a square spot of white upon his forehead, and on his back the figure of an eagle; the hairs in his tail are double, and there is a beetle upon his tongue."

Only this particular bull was worshiped in prehistoric Egypt! The old Egyptians perceived the sacred bull as a representative of the cosmos, a creation of the god Ptah. This very ancient cult worship is evidenced by the amulets of star-crowned bulls' heads found near Abydos or by the golden sun disks which were placed between the horns of the Apis bulls. Plutarch, a Greek historian and philosopher (around A.D. 50), wrote that the sacred bull was not born in a natural way; it arrived with a moonbeam falling out of the sky. This theory is supported by a stele which Auguste Mariette discovered in the Serapeum. The inscription above the Apis bull states, "You do not have a father; you have been created by the sky." According to Herodotus, "The Egyptians say that fire comes down from heaven upon the cow, which thereupon conceives Apis."

At some unknown, prehistoric point, the mysterious gods played a game with Apis (and other animals), but it was at a time "which we are unable to comprehend historically."[22] Thus the seeds of this animal cult were sown in mythical times, nurtured by the contradictory actions of the gods, which no human being could understand. These gods of supernatural origins performed impossible feats, deeds that simple humans were unable to grasp. They brought animals to life—who else can do that? They lived in animals and acted through animals. Animals provided the gods with information about mankind; they supported the gods in their struggles against man and against each other. Similarly divine is the creation of new animals, the coupling of species that did not interact naturally. All the hybrids, the monsters, and various sphinxes are divine creations. All of this was a little too much for the limited comprehension of simple people who had barely emerged from the Stone Age. Even the most creative and most daring human imagination needs a stimulus to work from. Nothing comes out of nothing—not even imaginary creations.

Animals on the Drawing Board

In my last book[23] I voiced a concern that has grown since then, and it is directly related to the animal cults of ancient civilizations. I was discussing the developments and further possibilities of gene technology. I pointed out that geneticists would soon be able to create new lifeforms and to combine existing ones. I wrote that development can be truly amazing and unexpected, and that it proves that changes in reality, in some instances can outstrip the boldest speculation. In April 1987, the U.S. Patent and Trademark Office announced that, in the future, it would grant patents for multicellular, living organisms if they are based on a program which does not occur in nature. Thus, the office legalized a development that had long been a reality: Up until March of 1987, more than two hundred genetically altered microbes had been submitted for patent rights—microbes which might neutralize oil spills or produce insulin, for example. In April 1987, fifteen patent applications were submitted for animals which did not exist in nature. Scientists at the University of California, for instance, succeeded in biotechnologically crossing a sheep with a goat; this new laboratory-generated creature had the front of a sheep and the rear of a goat. Horrified critics were supposed to be appeased by the assurance that this monster was only the prototype of a series and that the designers promised to improve future models.

I asked my readers: In the face of such evidence, who would dare deny the possibility that flying horses may have existed? Flying mice (bats) and flying fish have been a reality for thousands of years. I believe the question is justified whether such creatures are the products of natural evolution or the laboratory creations of alien visitors.

That was the state of affairs when I wrote in 1987. Since then, the clock has been ticking away.

In 1976, GENENTECH Corporation was founded in California. Its goal was to study the application of genetically developed drugs and to market them commercially. In its first years, the company had nothing but expenses for investments and wages; nobody was really sure it would succeed. Nowadays, its annual gross revenue exceeds $250 million. GENENTECH has been highly profitable for years, and its success has spawned a crop of about three hundred similar enterprises worldwide. What do they make? What perverted new outlet has capitalism found? In 1979, GENENTECH was able to duplicate the human insulin gene; a year later it developed a process to produce interferon-alpha. Shortly after that, GENENTECH created the genetically engineered drug Protropin, a hormone which corrects growth problems in children.

Licenses are issued for such products. Licenses bring money. In the near future, GENENTECH expects to receive a patent for a drug which will perform miracles in the healing of wounds. We are imitating the gods of ancient mythology: Open wounds are magically healed overnight.

On June 13, 1988, the German daily *Die Welt* reported that one of the most ambitious projects of molecular science, the complete decoding of the human genetic makeup, is taking on concrete forms. The "Genome Project," which has been hotly debated for over two years by scientists, is expected to cost three billion dollars. According to *Die Welt*, scientists are determined to analyze the entire human genetic material down to its smallest component within the next few years, at a tremendous cost in terms of personnel, equipment, and money.[24]

They will succeed. Genetically engineered humans are not too far off. The Genome Project is not aimed solely at humans, but is expected to include other organisms as well. After all, we are all related in some way, aren't we?

Geneticists at the University of Texas have developed a process which allows them to immediately distinguish genetically manipulated animals from the "true" or "original" animals. The process is very simple. The genetically manipulated genes are combined with an extra gene which triggers luciferase. That is the enzyme responsible for the cold light produced by lightning bugs. The enzyme is handed down to the following generations. All the descendants inherit the luciferase gene. A small tissue sample will be taken to determine whether an animal, several generations down the line, is descended from a genetically altered specimen—when treated with certain chemicals, the sample will start to glow.

I had always wondered how the mythological gods were able to distinguish certain creatures instantaneously from others of the same species. But the mystery is starting to unravel.

Dr. Tony Flint, director of the London Zoo, recently created an animal bank. Not for investments or financial trading, mind you. The animal bank collects eggs, sperm, embryos, and genetic material from animal species which are threatened with extinction. In this way, it is argued, geneticists of the future will have a chance to resurrect these species. Divine ambitions indeed!

Manetho and Eusebius—Two Witnesses

Is it really too far-fetched, too speculative, to apply our immediate future to our mystical past? Are the two completely unrelated, as some people suggest? Or could the very distinctive Apis bull have been the result of genetic manipulation? I would like to call to the stand two witnesses who are several thousand years older than me.

One of them is named Manetho. He was a high priest and scribe for the holy temples in Egypt. In the works of the Greek historian Plutarch, Manetho is mentioned as a contemporary of the first King Ptolemy (304–282 B.C.). Plutarch reports that the king ordered a heavy sculpture moved to Alexandria and that Manetho was the only person who was able to inform the king that the mysterious sculpted creature was a Serapis.[25] Manetho was an inhabitant of Sebennytus, a city located in the Nile delta, and that is where he wrote his three-volume book about the history of Egypt. He was an eyewitness to the fall of the three-thousand-year-old empire of the pharaohs; as a scholar he composed a chronicle about his gods and his kings. Manetho's original text has been lost, but essential passages from his work were adopted by the Greek historian Julius Africanus (who died in A.D. 240).

The second witness whose testimony I want to submit is also a historian. His name was Eusebius. He died in A.D. 339. He was bishop of Caesarea and one of the early Christian chroniclers. Eusebius relies heavily on Manetho, but includes many other sources as well. He indicates in the foreword to his *Chronological Tables* that he perused the manifold texts of such ancient writers as the Chaldeans, Assyrians, and Egyptians.[26]

Manetho and Eusebius often complement each other, although Eusebius tends to present the Christian interpretation while Manetho sticks to facts, numbers, and names. Manetho's account begins with a listing of gods and demigods and the times of their respective reigns. The dates are too much for our archaeologists to comprehend.[27] According to Manetho, the gods ruled Egypt for 13,900 years, followed by 11,000 years during which demigods ruled the country. (I will discuss this in detail later on.) Manetho writes that the gods created a number of creatures—monsters and all kinds of hybrids. Eusebius endorses this belief.

> And there were certain other creatures, some of whom had created themselves and were equipped with reproductive organs; and they created humans, with two wings; and others with four wings and two faces and one body and two heads, men and women, and two genders, male and female; furthermore, other humans with the thighs of goats and horns on their heads; and still others with the hooves of horses; and others with the body and legs of horses and the trunk and heads of humans, who resembled the hippocentaurs; moreover, they created bulls, with human heads, and dogs with four bodies, whose tails emerged from their hind quarters like fish tails; also horses with the heads of dogs; and humans and other monsters, with horses' heads and human bodies and fish tails; furthermore, all kinds of dragonlike monsters; and fishes and reptiles and snakes and a large number of wondrous creatures, in various shapes, all different from one another, whose images they kept in the temple of Belos, one next to the other.[28]

Hybrids Abound

That's some strong medicine this Eusebius wants us to swallow! One has to read his account two or three times, slowly, to register fully the preposterousness of his words. How was it back then? There were humans with two wings? Is all of this nonsense? If it is, why do we meet these creatures in hundreds of statues and sculptures displayed in all the major museums? They are not called "double-winged humans," but rather, with the total lack of imagination so common on modern archaeology, "winged geniuses." "Humans with goat thighs and horns on their heads"—are they really nothing but imaginary creatures? How about a look at some Sumerian and Assyrian seals and temple walls? There are hundreds of depictions of such mixed breeds. Similarly, we find a number of "horse-hoofed humans" and centaurs—half human, half horse—in ancient drawings. And the gods also created bulls with human heads. Remember the sacred Apis bull? The Louvre exhibits three small figurines, each about ten centimeters high, of bulls with human heads. They have been dated to 2200 B.C. (Let us not forget the monster of Crete, the Minotaur, a bull with a human head for whom the people of Crete built the famous maze.) Eusebius talks about dogs with fish tails, other monsters, and "a large number of wondrous creatures." And what about the sphinx! When we talk about the sphinx, everybody thinks of the huge statue of a human-faced lion next to the Great Pyramid of Giza. However, sphinxes come in all shapes and sizes. There are lion bodies with ram heads, dog or goat bodies with human heads, ram bodies with bird heads, human bodies with crocodile heads, and so forth. Entire zoos of different sphinxes have been dug from the desert sands or identified on Egyptian temple walls. Particularly strange creatures were discovered on the wall of a small side temple in Dendera in central Egypt. They combine the long-maned heads of lions or baboons with slender, almost human torsos and snake tails. These curious hybrids, dedicated to the goddess Hathor, rest elegantly on their curled-up tails. "Wondrous creatures," Eusebius calls them, "in various shapes, all different from one another."

Anybody who has taken the time to stroll through any large museum or to peruse a photographic book about Sumer, Assyria, or Egypt can confirm the abundance of such "wondrous creatures." The Museum of Baghdad features the statue of the "archaic goddess," a woman's torso with delicate breasts—and the head of a monster. A museum in Berlin displays a replica of the Ishtar temple gate in Babylonia. Scaly creatures with long tails and unusually long necks are parading on a blue, yellow, and brown enameled tile wall. Their front legs feature lion paws, while their hind legs are equipped with the sharp talons of eagles. The original is said to date back to 600 B.C. A Sumerian seal displayed at the Louvre in Paris and a tablet at the Egyptian Antiquities Museum in Cairo both show four-legged creatures with long, curved necks topped by snake heads.

Natural evolution never created such absurd hybrids. Are they merely an example of artistic liberty? Why are these beasts kept on a short leash by attending humans? Also at the Louvre, we can admire the twenty-three-centimeter cup of Gudea, which originates around 2200 B.C. The engraving on the cup shows a very unusual kind of hybrid with bird claws on its legs, the body of a snake, human hands, wings, and the head of a dragon. (Eusebius: "Furthermore, all kinds of dragonlike monsters . . .") A twenty-centimeter statue depicts a "winged goddess" with a graceful female torso, a girlish face, a woman's hands—in short, a perfect lady. The picture-perfect pinup loses much of its appeal, however, when we catch sight of the wings on her back and the repulsive animal claws on her feet.

There is certainly no shortage of artistic expressions of such "wondrous creatures." Hybrids and monsters abound at the Asutosh Museum in Calcutta, the Archaeological Museum of Ankara, the Museum of Delphi in Greece, as well as the Metropolitan Museum in New York, to name but a few.

A relief dedicated to the Assyrian king Assurbanipal (on view at the British Museum) shows a husky man leading a strange animal by a rope. It walks like a monkey on two feet, but its hands end in fins. The British Museum also owns a black obelisk made for the Assyrian king Salamasar II. Among other things, it depicts an elephant followed by two small creatures the size of children. They have human heads combined with the bodies, thighs, and legs of animals, and are kept on a leash by two guards. A different section of the same obelisk portrays two sphinxlike creatures with undeniably human heads. Nothing unusual, you say? Why does one of the sphinxes suck its thumb? Why are they chained up? Why does the text accompanying the pictures describe them as human animals being led into captivity?

Even in distant South and Central America, artists have immortalized hybrid creatures in their works. The Olmec, Mayan, and Aztec cultures feature numerous examples of human-animal monsters on temple walls and codices, always in connection with imperious gods. Eighteen years ago, I had the opportunity to photograph a number of metal tablets with indefinable creatures that were part of a collection belonging to Father Crespi of Cuenca, Ecuador. Father Crespi, who is now deceased, claimed to have received the tablets from local Indians. They are made from an Inca alloy of gold, copper, and zinc. In the summer of 1988, archaeologists made a most spectacular discovery near the small town of Sipan in northern Peru. They found the perfectly preserved tomb of a Moche priest and nobleman. (The culture of the Moche Indians was thriving along the Peruvian coast around the birth of Christ.) The nobleman had been buried in a wooden sarcophagus, accompanied by a large amount of jewelry, pearls, pottery, and gold, and was about thirty-five at the time of his death. The same family tomb contained four additional sarcophaguses with women and men. Several meters above the actual sepulcher, archaeologists encountered the skeleton of a man wrapped in cotton sheets. It was accompanied by a copper scepter, about one meter long, with a very explicit picture of a woman having sexual intercourse with a hybrid monster, half tomcat, half reptile.

In addition to these unequivocal visual reproductions, there are certainly a large number of hybrid creatures that no man or woman has ever seen. Many cultures describe the imaginary transformation of one monster into another. The image of a centaur, for example, may have resulted from the shadowy perception of a horse and its rider as being one. Similarly, the image of Pegasus may be rooted in wishful thinking, the human desire to own a winged horse.

In the adventures of Ulysses, the great Greek poet Homer (around 800 B.C.) mentions the sirens, who sing such a beautiful melody that the sailors forget their mission. Homer himself did not describe the sirens in any physical detail, but later writers turned them into winged women with birds' feet. Sirens have been a favorite subject of artists throughout the ages—although no artist ever laid eyes on one. The German legend of the Lorelei derives directly from the ancient sirens.

Hybrid creatures abound not only in the dramatic literature of ancient times, but also in more modern children's fairy tales. Around 700 B.C., the Greek writer Hesiod portrayed the monster Medusa, whose head sprouted a number of snakes, as being so terrible to behold that people would turn into stone upon seeing it. In his *Walpurgis Night*, the German dramatist Johann Wolfgang von Goethe imagines Adam's seductress as a snake with a woman's head. Elliot Smith sees the Chinese dragon as a hybrid between snake, crocodile, lion, and eagle.[29]

This and much more may grow from the multifaceted spectrum of human imagination without which there would be no fairy tales. I have to look beyond that. I am searching for a common denominator, the original key that opened the door for all these images to enter our thoughts and perceptions. After all, these wondrous creatures occur not only in the works of old Manetho and the theologian Eusebius, but also in Plutarch, Strabo, Plato, Tacitus, Diodorus Siculus, and Herodotus, although the latter is reluctant, as he states on several occasions, to describe such monsters.

Logically, there are only two possibile ways to explain the stories and artistic representations of these hybrid creatures:

1. They never existed. They are, without exception, the product of someone's imagination. In that case, the

painters, sculptors, and writers must have exaggerated wildly.

2. Such hybrid creatures did in fact exist at some point. If that is the case, the "wondrous creatures" of Eusebius could only have evolved through genetic design. Any other conclusion is impossible, because natural evolution cannot produce such hybrid monsters. The reproductive organs and chromosomes of the animals involved are too different. Sexual interaction would not produce any offspring. Does this make sense?

I cannot walk through a rain shower without getting wet. In dealing with these "wondrous creatures," I walked through some heavy showers and got wet to the bone. But my thoughts remained firmly grounded in reality. I admit that the idea that these monsters, as they are described by Eusebius and others ("in various shapes, all different from one another"), could actually have lived seems far-fetched at first. Our minds, so comfortably accustomed to the cuddly creatures we find in nature, refuse to consider a zoo of living monsters. I will be accused of trying to make reality conform to my own wishful thinking. And I will consider myself in good company, looking at the ancient writers. Is the very idea that "wondrous creatures" may have existed proof that they could not have existed? Are stories wrong just because they are legends? Who turned the ancient stories into legends? Could it have been our limited understanding? Could it have been the narrow and inflexible horizon of academic rationale that dictates to each generation how far its thinking may go? I suspect that much of what we dismiss as unrealistic and irrational was once living history. The

Roman philosopher Lucius Apuleius, who lived in the second century A.D. and who traveled in Egypt, wrote in *The Metamorphoses*: "O Egypt! Egypt! Your knowledge will survive but in legends, which later generations will be unable to believe." The following fable from the eternally young genre of science fiction could shed some light on our problem here.

A Science-Fiction Model

There once was a time when the gods ruled the earth. Man did not know who the gods were or where they had come from. Barely evolved past the animal stage, humans passively waited for the light. The gods lived in the sky, somewhere up there among the stars.

There, among the asteroids between Jupiter and Mars, aliens had anchored their mother ship. The long space flight between the stars had used up most of their energy supply; it was time for a pit stop to secure further resources and perform the necessary maintenance. Thus, the gods were forced to spend a couple of centuries in our solar system. The years passed slowly; the gods got bored quickly. In search of variety and entertainment, they devised games and contests. They had no concept of human morality or ethics in today's sense of the word. Their feeling and thinking came from a different dimension. Earth was their playground.

One day, Ptah, divine designer of organs, created a new creature on his drawing board. The genetic material came from two stupid life forms on earth. The ensuing combination between a lion and a sheep spawned a new vegetarian creature with the paws and the speed of a lion.

But Ptah was shocked to discover that a real lion had no trouble tearing the divine creature to shreds. That was certainly unacceptable! Khnum said to Ptah, "The brain of the sheep was inferior to that of the lion. Try a different combination, with a lion torso and a bull's head." This monster survived because the lions of the world avoided it.

Ptah was all ready to celebrate his new creation when something incredible happened. The primitive two-legged creatures on earth banded together and killed the monster with their spears and slingshots. Like lightning, Ptah descended on mankind, angrily punishing the clumsy humans, who did not understand his design.

The council of the gods reprimanded Ptah for his actions. "It is wrong," they said, "to punish people for their behavior if they have not been warned in advance of the negative consequences." Ptah agreed and began to mark all his new creations with a visible sign: a light-colored rectangle on the forehead, for instance, or two shining horns on the temple. Now the humans knew which creatures were the gods' property and which they were allowed to kill for food. The extraterrestrials had found a way to keep themselves busy. They merrily invented one monster after another, "in various shapes, all different from one another." They studied the behavior of their creations, their usefulness, their interaction in a natural environment, and they observed with much amusement the reactions of the flabbergasted humans.

Finally, the mother ship was replenished with the necessary resources. It was time to explore some new territories of the universe. The gods left behind a subdued human race with all the animals that had populated the earth since the beginning—plus the divinely created monsters. The priests were the first to realize that the gods had vanished. Scared and unsure, they did not dare to lay a hand on the gods' new creations. Generations came and went. Many of the divine beasts died and became extinct; but others, those with "life-producing forms" (Eusebius), mutated, were taken captive, or were pampered in the temples. The priests kept the knowledge alive of certain creatures dedicated to the gods. And because the priests feared that the gods could return at any given moment without notice, they watched the sky apprehensively for any unusual movement. Novices were commissioned to search the land for the gods' sacred animals and to bring them to the temples so that they could be properly worshiped. It was obvious to the priests that the sacred animals, if they died, had to be mummified and buried in splendor; after all, they belonged to the gods, whose return might be imminent.

Centuries passed, even millennia. Times changed, and so did people. In folk tales, the memory of terrible monsters lived on. Nobody was afraid of the small creatures, like birds, fish, or pets. People were able to talk to them, maybe use them to intercede with the gods on their behalf. But what about the large, fearsome beasts? Would they be changed back into their horrible original form after death? Would they frighten and terrorize people if they were to be reborn? What could man do to satisfy the gods without suffering under these beasts?

The priests pondered this momentous question for a very long time. Finally, they found a simple solution. As long as they were alive, the animals were to be pampered, worshiped, idolized, so that their ka and ba would ascend to the gods after their death, where they would be able to attest to the benevolence and respect with which mankind had treated them. The bones of the animals, however, were to be fractured, smashed into pieces, and mixed with asphalt after their death. Their remains were to be placed in heavy sarcophaguses made from the hardest granite, coffins so massive and huge that no monster, if it happened to be reborn, would be able to escape from them. The sarcophaguses had to be anchored deep in an underground vault. These monsters should never be able to attack and tyrannize people again.

Pseudo-Animals in the Wrong Vaults

We need models, new ideas, to make even marginal sense of the absurdities and contradictions that our ancestors left us with. The story I invented here cannot be considered anything more than such a model—or a crutch, if you prefer, to help us hobble out of the swamp of our own past. We are more than willing to accept and appreciate the writings of Herodotus, Strabo, Diodorus Siculus, Tacitus, Manetho, or Eusebius if they fit our common pattern of knowledge. But heaven forbid that they not fit! Arrogantly, we elect ourselves judge and jury and categorically dismiss the same writings of the very same authors. Even if it is obviously the truth, it cannot be.

What did Auguste Mariette find on September 5, 1852, in the undamaged bull sarcophaguses of Saqqâra?

As he reported, his first concern was for the head of the bull, but he found none. The sarcophagus contained a foul-smelling bituminous mass which disintegrated under the slightest pressure.

Were the bones of this pseudo-animal smashed many years or centuries *after* the actual burial? Not according to Mariette. He said that the stinky mass was permeated with a large number of tiny bones which apparently had been fractured *at the time of burial*.

And what about the second intact sarcophagus? According to Mariette, there were no skulls of bulls, no large bones; on the contrary, he found an even greater supply of tiny bone fragments.

Why did the archaeologist Sir Robert Mond find bones in bull sarcophaguses which he *assumed* were those of jackals or dogs? I do not blame an anthropologist for not examining such bones any further. Why should it occur to him or her that there may have been dogs with four bodies whose tails issued from their backs in the manner of fish tails (Eusebius)?

Dr. Ange-Pierre Leca is a physician and specialist in Egyptian mummies, and is the author of an exciting book on the subject.[30] In it, he mentions two "wonderfully bandaged bulls" with "a beautiful outside" which were found in the vaults of Abu Sir. Inside the second mummy, which appeared to be a single bull, Leca found the bones of seven animals, among them a two-year-old calf and a huge old bull. A third looked as if it had had two skulls.

I beg your pardon? *Two skulls?* Just ask Eusebius, who described "wondrous creatures, in various shapes, all different from one another . . . with one body and two heads."

Naturally, I informed Dr. Holeil Ghaly, the chief excavator of Saqqâra, of my suspicions. I asked him if he or any of his colleagues had ever found mummies of animals whose bones did not seem to go together at all. He stared at me pensively—and a little incredulously, I thought. "My God, who would pay attention to something like that?" is what I imagined him saying to himself.

No one. The very thought is absurd.

The zealous excavator Walter Emery discovered catacombs of sacred cows in Saqqâra. There was no doubt about their purpose, because the inscriptions on the carefully hewn limestone blocks read: "Here rests Isis, mother of Apis." Moreover, the excavators found several well-preserved papyri from the third and fourth centuries B.C. which invoke and praise the cow goddess. But instead of a cow mummy, the archaeologists encountered a bundle of wrapped bones from calves and other animals. Jean Philippe Lauer, Emery's successor, identified the bones, "beyond a doubt," as being from plundered graves. Yet he could not find an entrance to any such graves.[31]

As I mentioned before, grave robbers are interested in objects of economic value. They mess up and destroy the graves. They are not academicians, or pranksters. It is difficult to imagine why grave robbers would have transferred bones from a different tomb into the vaults of the sacred cows.

The Mystery of the Baby Baboon

A baboon with the head of a dog lived in ancient Egypt and Nubia (now Sudan). The Egyptians considered it sacred. This baboon was even part of the tribute which the Egyptians exacted from the Nubians. Thousands of these funny-looking creatures with the jaws of a dog and a thick mane were mummified. Nobody has ever thought much about it; after all, similar baboons exist today. But there is one curious discovery that deserves some further scientific study.

In 1972, Dr. Henry Riad, then director of the Egyptian Antiquities Museum in Cairo, granted permission to several scientists to x-ray and examine mummies. Professor James E. Harris from the University of Michigan studied the mummy of the priestess Makare. She held the highest title in the female hierarchy, as "wife to the god Amun."[32] The way she was bandaged indicates she died in childbirth, because the body of the baby, also wrapped in bandages, was placed on the mother's body in the sarcophagus. The tiny bundle was x-rayed from all angles, and the scientists were speechless. What had been thought to be a baby turned out to be a dog-headed baboon with a slightly enlarged brain volume!

I believe it is justifiable to ask whether this woman, priestess to the god Amun, gave birth to the little monster. Herodotus makes no secret of his disapproval regarding the sexual perversion of the Egyptian priests. He maintains that the Egyptian painters and sculptors depict the god Pan with the head and the legs of a goat. "They represent him thus for a reason which I prefer not to relate." But a few lines later he remarks angrily that "a he-goat had intercourse with a woman publicly." Apparently, Diodorus Siculus also knew more than he let on when he wrote that the origins of animal worship had to be kept secret.

The few Egyptologists I know are open-minded people who have contributed greatly to the solution and reconstruction of the ancient Egyptian mysteries. Egyptology has a unique position within the field of archaeology. Egypt is the only place where scores of temples and sculptures have been reclaimed from the desert sands over the past decades, thanks to the tenacity and infinite diligence of Egyptologists. As hieroglyphs have been deciphered, the history of ancient Egypt has become more transparent. Egyptologists know what they are talking about. They will accuse me of hiding the fact that some genuine mummified Apis bulls were found as well. They can be admired at the Louvre or at the Museum of Natural History in Vienna, Munich, and New York. I am well aware of their existence, but I am also aware of the fact that their origins and contents are highly obscure. Anybody who has studied this topic at length knows that the priest Manetho, of all people, was a staunch supporter of the Serapis cult and that real bulls undoubtedly were buried during his lifetime. We also know of the texts honoring Apis that were found in the Serapeum of Alexandria (and in other places).[33] But all of this dates back to the Ptolemies and the Romans, a mere two thousand or two thousand five hundred years before our time. I am not aiming at this recent era; I am aiming at the origins of the animal cults, which go back into ancient history. Isn't it odd that the first Pharaoh Ptolemy (304–284 B.C.) has a heavy sculpture hauled back to Alexandria which has been lying around in the dirt, and no one knows what the sculpture represents? The priest Manetho, who happens to be nearby, is the only one who can enlighten his king on the subject. He announces that the mysterious figure is a Serapis. ("Serapis" is the Greek word for the sacred bull.)

This brief anecdote, related by Plutarch, leads to an ominous conclusion: The king and his entourage were all stupid. They did not even recognize the sculpture of a bull. I wonder why not. Probably because the sculpture showed one of the "wondrous creatures." Only a priest like Manetho would have been able to explain that.

Dostoyevsky was right: Nothing is harder to believe than reality.

◇ 2 ◇

The Lost Labyrinth

"One must not mistake majority for truth."

JEAN COCTEAU,
FRENCH POET (1889–1963)

Ten years ago, I thought it would be pointless to write about Egypt. Indubitably, we know all there is to know—don't we? Rather reluctantly, I was leafing through books about Egypt. Pyramids, and more pyramids! Sphinxes! Pharaohs! Strange gods with even stranger headdresses—downright comical. Since my job entailed a lot of visits to museums all over the world, I kept running into these ancient Egyptians wherever I went. Gradually, I became familiar with the names of the various gods, and soon I greeted them like old friends whenever I saw them on their pedestals or in their glass cases at the museum. Hathor? Oh, she is the one who gracefully balances the horns of a cow and a sun disk on her hair. Thoth? I know. He's the strapping young man with the face of a bird and the half-moon and sphere over his head. He's an old colleague by the way, because Thot is the god of writers. Sobek? Isn't he the weird one with the crocodile's head that has antennas attached to it? Min? No mistaking him, with that double row of sun batteries over his hood. Clearly, he needs energy: He's always shown with a three-tailed whip. Horus? My old buddy. No one visiting Egypt could possibly miss him. His tool, the winged sun disk, beckons from almost every gilded ceiling and dominates the monumental entrances to many temples. A very suitable logo for flying trains (magnetic trains) or UFOs. Add to that the eye of Horus, always watchful, always present above the earth, singularly suited to be the god of satellite builders. Horus himself is depicted in different ways in Upper versus Lower Egypt, either as a human with a falcon's head or as a full-fledged falcon. His double crown reminds me, irreverently, of a ladle with a liquor bottle in it. (Oh well, we have to relate somehow to these ancient headdresses.)

There are myriads of gods, male and female; some are human-animal hybrids, others plain animals. The heavenly world of the Egyptian gods is rather simple until we start looking at the family ties between the different gods; moreover, creative human storytelling is bound to exaggerate or distort the "facts" about these divine creatures. Why should Egypt be any different than ancient Greece, India, Japan, or Central America? Mankind needs a god for every little problem, as evidenced by the various Christian saints canonized in the last two thousand years.

At some point, I got hold of a book about the Egyptian gods. It was dry and monotonous reading. I remember how grudgingly I struggled through the book. I could not have cared less about which god fathered which divine offspring or which incest produced which godly brat. I figured I would be able to look up this information in any of a number of excellent encyclopedias about mythology if I should ever need it. Besides, archaeologists have done a marvelous job of listing all the names and reigns of all the pharaohs; they have labeled every temple and every statue and have discussed every piece of artwork in great detail. It was very doubtful that I would ever write a book about Egypt. It just was not my cup of tea! I am a detective, an explorer in search of unsolved mysteries—what mysteries did Egypt have left that needed to be solved?

From Saul to Paul

My attitude changed abruptly a few years ago when I was looking for some unrelated information in the works of Herodotus. By Jove! Herodotus was telling stories that did not coincide at all with the information promulgated by Egyptologists. Who was right, the historian of twenty-five hundred years ago or the modern archaeologists? Was Herodotus an individualist given to exaggeration or did other eyewitnesses from his era corroborate his accounts? The discrepancies between Herodotus and today's interpretation of events were often so blatant and shocking that I began to investigate. The further I plunged into the ancient literature, the more fascinating Egypt became! I was hooked! This could not be true! Had those Egyptologists whom I had praised so highly been asleep at the wheel? Were they satisfied with simply gluing the pieces of the puzzle back in place, superficially, without investigating the discrepancies underneath? Was I on the trail of some ancient knowledge that only the dignitaries of obscure secret organizations were privy to? Were there things to be learned about ancient Egypt that did not fit into our modern world, that were not opportune, that were better left unspoken, so that average people would not be scattered in panic like a flock of birds?

Three thousand years ago, the Phoenician historian Sanchuniathon (around 1250 B.C.) must have pondered the same question when he wrote that "from early youth, we are accustomed to hearing falsified reports, and our minds have been saturated with prejudice for centuries, to the extent that we guard the fantastic lies like a treasure, so that in the end the truth becomes unbelievable and the lie appears to be true."

The philosopher Cicero (106–43 B.C.) declared Herodotus the "Father of History." Although he certainly was not the first historian, Herodotus has kept the title to this day.

Who Was Herodotus?

What do we know about Herodotus? He was originally from Halicarnassus, a town in the southwest corner of Asia Minor. His father rebelled so vehemently against the tyrant ruler Lygdamis that the whole family was banished into exile. Things were no different then than they are today. From the island of Samos, Herodotus observed the political events of his world. It was not a quiet time. Greece was threatened by the powerful Persian empire. Athens had initiated the first Attican naval alliance and was competing against its military rival, Sparta. Maybe the political dissension prompted young Herodotus to probe and investigate matters personally. He became the premier globe-trotting journalist of his time. He traveled all over Asia Minor, Italy, Sicily, southern Russia, Cyprus, Syria, and even as far as Babylonia, where he stayed for some time. When Herodotus entered Egypt in July 448 B.C., it was no *terra incognita*, no unexplored territory. One of his countrymen, a natural philosopher named Hecataeus (around 550–480 B.C.), had already described the land on the Nile. Herodotus, however, did not accept his predecessor's judgment uncritically; on the contrary, he viewed the writings of Hecataeus with a certain prejudice and a strong distrust.[1]

Herodotus never acted exclusively as a historian. He scrupulously noted all that he was told about the history of a country, but he also described the geography and topography of the areas he visited. He was just as much geographer as he was historian. Herodotus was the first to express the notion that all history must be viewed in the context of its geographical location and that every geographical area has its own history.[2]

At the time, Egypt had established strong commercial ties with Greece. The Persian king Artaxerxes I (465–424 B.C.), who ruled the land on the Nile, used to send Egyptian boys to Greece for language lessons. Conversely, Greek merchants and restaurant owners had businesses in Egypt. Herodotus, who spoke no Egyptian, had to rely on interpreters, but there was no shortage of those. His sources were priests of various ranks from the sacred temples in Memphis, Heliopolis, and Thebes, as well as librarians, some members of the king's court, and Egyptian noblemen, who enjoyed conversing with the Greek visitor.

Herodotus quickly perceived the distinction between

popular traditions and the official history of Egypt contained in the papyri of libraries and temples. When one of the priests read to him the names of 331 pharaohs, Herodotus wrote them down exactly, but when someone told him a story about the cow of Menkaure, he dismissed it as idle talk. He scrupulously recorded the heroic deeds of ancient pharaohs, but he immediately became suspicious and sceptical when he heard folk tales such as the one claiming that sixteen hundred silver talents had been spent on radishes and onions during the construction of the pyramid.

An astute listener, Herodotus did not just write down gullibly and in utter amazement whatever was brought to his ears; he often added his own acrid comments. He supplemented what he heard with his own reporting, carefully separating secondhand accounts from his own interpretations. The following is an eyewitness report drafted twenty-five hundred years ago.[3]

Bigger Than the Pyramid?

> To bind themselves yet more closely together, it seemed good to them [the kings] to leave a common monument. In pursuance of this resolution they made the Labyrinth which lies a little above Lake Moeris, in the neighborhood of the place called the city of Crocodiles. I visited this place, and found it to surpass description; for if all the walls and other great works of the Greeks could be put together in one, they would not equal, either for labour or expense, this Labyrinth; and yet the temple of Ephesus is a building worthy of note, and so is the temple of Samos. The pyramids likewise surpass description, and are severally equal to a number of the greatest works of the Greeks, but the Labyrinth sur-

passes the pyramids. It has twelve courts, all of them roofed, with gates exactly opposite one another, six looking to the north, and six to the south. A single wall surrounds the entire building. There are two different sorts of chambers throughout—half under ground, half above ground, the latter built upon the former; the whole number of these chambers is three thousand, fifteen hundred of each kind. The upper chambers I myself passed through and saw, and what I say concerning them is from my own observation; of the underground chambers I can only speak from report: for the keepers of the building could not be got to show them, since they contained (as they said) the sepulchres of the kings who built the Labyrinth, and also of the sacred crocodiles. Thus it is from hearsay only that I can speak of the lower chambers. The upper chambers, however, I saw with my own eyes, and found them to excel all other human productions. . . . At the corner of the Labyrinth stands a pyramid, forty fathoms high, with large figures engraved on it, which is entered by a subterranean passage.

Wonderful as is the Labyrinth, the work called the Lake of Moeris, which is close by the Labyrinth, is yet more astonishing. . . . It is manifestly an artificial excavation, for nearly in the centre there stand two pyramids, rising to the height of fifty fathoms above the surface of the water, and extending as far beneath, crowned each of them with a colossal statue sitting upon a throne.

The Great Pyramid of Giza is arguably the most monumental building in Egyptian history, one of the seven wonders of the world. How can Herodotus, who knew this pyramid very well, describe a maze that is beyond words and even surpasses the Great Pyramid? Did the hot Egyptian sun bake his brain? Not likely, considering the fact that Herodotus emphasizes four times in the above passage that he actually saw this labyrinth: "I visited this place, and found it to surpass description"; "the upper chambers I myself passed through and saw, and what I say concerning them is from my own observation"; "the upper chambers, however, I saw with my own eyes"; and "it is manifestly an artificial excavation." I am grateful for his distance, his clear distinction between the things he marvels at with his own eyes and the things he is told by others: "of the underground chambers I can only speak from report" and "thus it is from hearsay only that I can speak of the lower chambers."

If we are to belive Herodotus, the Labyrinth must have been a fabulous construction achievement. Imagine fifteen hundred rooms aboveground, with covered yards and one wall surrounding the whole humongous complex. Gigantic! As if that were not enough, there is also a large artificial lake from which two pyramids protrude.

I have to stretch my imagination to the limit to come up with even a faint idea of how such a gigantic building could have disappeared from the face of the earth without a trace. Clearly, it was there in the fall of 448 B.C. It could be argued that later generations took the maze complex apart, piece by piece, to use as construction material for other buildings. But who could have done it? During the lifetime of Herodotus and thereafter, Egypt did not produce another single outstanding edifice. The pyramid-building era had long passed; the temples were crumbling. The Romans, Christians, and Arabs who succeeded the ancient Egyptians did not build anything out of the ordinary. Then again, does it have to have been something extraordinary? After all, the ancient towering monuments could have been dismantled and used piecemeal for building houses and roads. There are a number of proven incidents worldwide where this was the case. But where are these Egyptian stone houses erected with enormous blocks from the ancient Labyrinth? Where are these extravagant highways paved with ornate stones with lavishly detailed engravings of the gods? Herodotus describes the interior of this miraculous building.

A World of Wonders

> For the passages through the houses, and the varied windings of the paths across the courts excited in me infinite admiration as I passed from the courts into chambers, and from the chambers into colonnades, and from colonnades into fresh houses, and again from these into courts unseen before. The roof was throughout of stone, like the walls; and the walls were carved all over with figures; every court was surrounded with a colonnade which was built of white stones exquisitely fitted together.

The Egyptians never used such luxurious artworks—walls "carved all over with figures"—for building houses or streets. Even during the reign of the Ptolemies and the Romans, religious traditions were highly regarded. The Romans were no barbarians. Their historians would have recorded the pillaging and dismantling of such a unique and beautiful edifice. But the records make no mention of anything like that. Did Moslems disassemble the maze? Did they use the materials to build mosques or the huge citadel in Cairo? The center of modern-day Cairo developed from an army camp in the middle of the seventh century. None of the building blocks used were lavishly decorated or unusually large. When Sultan Saladin ordered the construction of the impressive citadel in 1176, no one even remembered the existence of a spectacular ancient labyrinth. There is more to consider than simply the dismantling of a unique monument (surpassing even the pyramids, according to Herodotus). Consider the transportation of these monumental granite blocks, marble columns, and colossal stone sculptures which Herodotus describes. Such extraordinary transportation exploits, with all the attending logistics problems, were realized during the early times and the pinnacle of the pharaonic empire, and never again after that!

Was the maze swallowed by the desert sand? Possibly. The sand engulfed entire pyramids and even the giant sphinx of Giza. But where, for heaven's sake, are the fifteen hundred rooms *below*ground, with the sepulchers of the twelve kings? I have often been accused of having a lively imagination. I can visualize a fairy-tale palace from the *Arabian Nights*. What kind of superhuman effort was required to construct fifteen hundred chambers below the ground? The ancient tunnel builders had no dynamite or modern drilling equipment at their disposal. I would assume that the fifteen hundred underground rooms were luxuriously decorated and filled with reliefs and sculptures. After all, they housed the tombs of twelve kings. What would have been used to light so many subterranean chambers? What kind of ventilation system would have been required during construction? What kinds of pictures or writings adorned the walls? How deep were the sarcophaguses of the twelve kings? What messages from an ancient culture are waiting to be deciphered in these fifteen hundred rooms?

Holy Osiris! Shouldn't every Egyptologist be drooling

at the prospect of finding this labyrinth? Where else in the world are we likely to find something like this? Even if the fifteen hundred aboveground rooms which Herodotus described are no longer in existence, we should be able to locate the ruins of the gigantic wall, the foundations for the large colonnades, and maybe the crossbeams over the massive gates. Logically, the underground vaults would be easy to find after that. Ever since I started to delve into Herodotus' historical account, I have been wondering why no archaeologist has taken steps to search for the sepulchers of the twelve kings. Why do they just shrug it off? How can they be so indifferent in the face of such a sensational prospect?

A Matter of Faith?

I know the reasons behind their indifferent and lazy attitude. Some archaeologists claim that Herodotus' account is not believable. The majority, however, are convinced that the maze has long been discovered.

Much has been written about Herodotus' credibility—not just some articles in academic journals, but full-length books, thick and voluminous. I concede to every scholar studying Herodotus genuine effort and integrity, but every evaluation of his works ultimately remains subjective in nature, since none of us knew him personally. We can draw but indirect conclusions about his character. Was he strong-willed? Temperamental? Easygoing? A quiet listener and note-taker? Among scholars, the Father of History is interpreted as a studious collector of information, a pleasant storyteller, a semiscientific amateur, or even a liar. He is praised for his extraordinary memory and criticized for arrogance.[4] In 1926, the German philosopher Dr. Wilhelm Spiegelberg maintained that Herodotus reported the Egyptian legends the way he heard them and that he can be completely trusted in this respect.[5] O. Armayor Kimball, on the other hand, concluded in 1985 that the labyrinth Herodotus described is "out of the question in the real world of Egyptian history."[6]

The geographer Hanno Beck is a bit more lenient in his assessment of Herodotus' work. He suggests that it was unavoidable that Herodotus misinterpreted some aspects, or that exaggerations or mistakes found their way into the Greek writer's works, because Herodotus did not know or speak the language of the people he was visiting. Nevertheless, Beck applauds him for his attempts to critique the information he received.[7]

Friedrich Oertel, finally, the author of a well-researched evaluation of Herodotus, concluded that nothing emerges from the historian's description of Lower Egypt that would impugn his credibility. On the contrary, Oertel maintains.[8]

In studying several rather profound works for and against Herodotus, I realized that the negative conclusions were always a reflection on the scholars who expressed them. They base their judgment on today's state of knowledge. Because this or that has not endured *to the present day*, because they cannot imagine this or that *at the prersent time*, they conclude that Herodotus must have been wrong. But how well does our present knowledge really hold up against five thousand years of history? There is a Chinese saying: All people are smart; some before, some after.

Herodotus did not invent the maze and its adjacent artificial lake. Diodorus Siculus also described it in the first century B.C.[9]

Eyewitness Report

> After King Actisanes died, the Egyptians regained the sovereignty for themselves and chose as king their countryman Mendes, whom some also call Marrus. This king performed no martial deeds whatever, but he did build himself a place of burial known as the Labyrinth, a work remarkable not so much for its size as for its unrivaled cleverness of construction: for it is hard for anyone venturing inside to find his way out again, unless he has obtained a guide of fully proven experience. And some say also that Daedalos, who during a visit to Egypt had marveled at the ingenuity of the building, made for King Minos of Crete a labyrinth similar to the one in Egypt, and in which, say the myths, lived the so-called Minotaur. But the one in Crete has wholly disappeared, either because one of the rulers demolished it or because time has destroyed the work; while the one in Egypt has been preserved in its entirety down until our lifetime.

Diodorus also repeats Herodotus' story about the twelve kings and the sepulcher they shared. He confirms the fact that the maze is located at the place where the canal flows into Lake Moeris. Diodorus claims that the artwork is so superior that it cannot be surpassed.

Four hundred and twenty-three years after Herodotus, another eyewitness visited the same spot: the Greek geographer Strabo (around 63 B.C.–A.D. 26) traveled extensively, and in 25 B.C. his journeys led him to Egypt. Strabo's historical accounts have been lost, but most of his seventeen-volume *Geography* has survived. In the last volume,[10] Strabo wrote:

> Be this as it may, the Lake of Moeris, on account of its size and its depth, is sufficient to bear the flood-tides at the risings of the Nile. . . . locks have been placed at both mouths of the canal, by which the engineers regulate both the inflow and outflow of the water. In addition to the things mentioned, this Nome has the Labyrinth, which is a work comparable to the pyramids, and, near it, the tomb of the king who built the Labyrinth. . . . for this is the number of courts, surrounded by colonnades, continuous with one another, all in a single row and along one wall. . . . In front of the entrances are crypts, as it were, which are long and numerous and have winding passages communicating with one another, so that no stranger can find his way either into any court or out of it without a guide. But the marvellous thing is that the roof of each of the chambers consists of a single stone, and that the breadths of the crypts are likewise roofed with single slabs of surpassing size, with no intermixture anywhere of timber or of any other material. And, on ascending to the roof, which is at no great height . . . one can see a plain of stone, consisting of stones of that great size . . . . and their walls, also, are composed of stones that are no smaller in size. At the end of this building . . . is the tomb, a quadrangular pyramid, which has sides about four plethra in width and a height equal thereto. Imandes is the name of the man buried there. . . . Sailing along shore for a distance of one hundred stadia, one comes to the city of Arsinoe, which in earlier times was called Crocodeilonpolis. . . . our host, one of the officials, who was introducing us into the mysteries there, went with us to the lake. . . .

Like Herodotus, Strabo was deeply impressed by the size of the maze and its huge slabs of stone. But conspicuously, Strabo made no mention of the fifteen hundred chambers below the ground. Why not? Strabo's visit took place during the Roman domination of Egypt. In the year 47 B.C., the Roman Emperor Gaius Julius Caesar (100–44 B.C.) defeated the Egyptian army and installed his mistress Cleopatra as queen of Egypt. Seventeen years later—five years prior to Strabo's visit—Egypt was declared a Roman province. Obviously, the Egyptian priests had no intention of revealing their ancient secrets to the Roman conquerors. In Egypt, as elsewhere, Julius Caesar and his army were known for their ruthless plundering. It is likely that the Egyptian priests did what their colleagues in Central and South America did after the Spanish conquerors arrived: They hid their cultural treasures. Even 423 years before Strabo, Herodotus was not permitted to enter the subterranean halls. It does not seem surprising, then, that Strabo did not even hear about the underground tombs. Although he was Greek, Strabo was a subject of the dreaded Roman Empire because Greece, at the time, was part of that empire.

Moreover—and this cannot be overemphasized—almost five hundred years separate the visits of Herodotus and Strabo. In comparison: Construction on the cathedral of Cologne, Germany, was begun in 1248. Two hundred years later, the south steeple only ascended to the belfry. The cathedral was not completed until 1880. Five hundred years ago, the architects and builders, in all likelihood, would have been able to tell us something about the catacombs under the cathedral. Today, the tourists never hear about them. The time frame between Herodotus and Strabo is 423 years! That is no jiffy! At the time of Herodotus' visit, the priests were able to inform him proudly that he was seeing but half of the Labyrinth, and that the other half, just as impressive, was buried under the ground. When Strabo came to call, the priests either no longer knew anything about the underground chambers or kept quiet for political reasons. It is possible that Strabo heard rumors about the royal sepulchers below the maze but did not believe them and therefore chose to exclude them from his writings.

One hundred years after Strabo, the Roman historian Gaius Plinius Secundus (A.D. 61–113) also described the Egyptian Labyrinth. Once more, we discover details that none of the previous authors mentioned. Apparently, Plinius had sources that neither Herodotus nor Strabo had access to; interestingly, Plinius refers to Herodotus in an attempt to correct and supplement the latter.[11]

> . . . let us enter also into the labyrinths; which we may truly say, are the most monstrous works that ever were devised by the hand of man: neither are they incredible and fabulous as peradventure it may be supposed, for one of them remaineth to be seen at this day within the jurisdiction of Heracleopolis, the first that was ever made, to wit, three thousand and six hundred years ago, by a king named Petesuccas, or as some think Titoes: and yet Herodotus saith, it was the whole work of many kings one after another, and that Psammerichus was the last that put his hand to it and made an end thereof. The reason that moved these princes to make this labyrinth, is not resolved by writers, but divers causes are by them alleged. . . . Certes, there is no doubt made that Daedalus took from hence the pattern and platform of his labyrinth which he made in Crete; but surely he expressed not above the hundredth part thereof.

Plinius points out that the labyrinth of Daedalus was the second one; a third one was found in Lemnos, a fourth one in Italy. All were arched with polished stones; the Egyptian Labyrinth consisted mostly of large blocks of syenite. Even the centuries could not destroy it, although the people of Heracleopolis did their best to damage the building, which they despised. According to Plinius, the Labyrinth held the temples of all the Egyptian deities; goddesses of vengeance in forty chapels; and several pyramids, each forty yards high, whose bases covered an area of six acres. Already tired from walking, Plinius entered those convoluted, crisscrossing passageways. But there were also dining halls above, and ninety steps descended into the galleries below, which featured porphyry columns, religious idols, statues of the kings, and all kinds of horrible creatures. Some of the houses were situated in such a way that a terrible thunder echoes through them each time the door was opened. Plinius affirms that in most of the Labyrinth one was surrounded by darkness. Outside the wall, there were several other large buildings, called wings; moreover, some houses were located in underground vaults.

Of all the ancient accounts, Herodotus' description contains the most detail. This seems logical since Herodotus was the first one, of all the authors discussed here, to visit the maze. His eyewitness report and his account of what the priests told him about the underground structures date back the farthest in history; in other words, they are closest to the distant reality.

Even if the various writers ascribe different names to the builders of the Labyrinth, they agree on the main points. The mazelike temple structure is located on the shores of Lake Moeris, there are artificial canals, and the city of crocodiles is nearby. The aboveground buildings are a superhuman accomplishment, the ceilings are made of stone, the walls are also constructed with slabs of extraordinary size, and from the low roof "one can see a plain of stone, consisting of stones of that great size" (Strabo). No wood was used in the construction, according to Plinius and Strabo, but all three writers report that very close to the maze stood one or more pyramids. Both Herodotus and Plinius mention subterranean halls, and Herodotus and Diodorus also describe two additional pyramids emerging from the artificial lake. Only Plinius, however, tells of religious idols and all kinds of horrible creatures.

What has happened to this mythical maze, so elaborately praised by ancient historians?

A majority of Egyptologists believe that the Labyrinth was discovered in 1843 by the famous German archaeologist Richard Lepsius (1810–1884). It is thought to be identical to the burial pyramid of Pharaoh Amenemhet III (1844–1797 B.C.) and its surrounding ruins, which Lepsius located near the modern-day oasis of Fayyum.

The Merry Archaeologist

Is this assumption correct? What made Lepsius conclude that he had indeed discovered the Labyrinth? Did he enter the fifteen hundred rooms? Did he view the sepulchers of the twelve kings? Did Lepsius and his crew from the Royal Prussian Egypt Expedition indeed find stone slabs of extraordinary size or entire stone plains made of equally large slabs (Herodotus)? Did the excavators run into all kinds of horrible creatures (Plinius) or winding, convoluted paths (Strabo)?

They found none of the above!

In the nineteenth century, Richard Lepsius, the son of a district president from Naumburg, was considered the undisputed genius among German archaeologists. He was eccentric, obsessed, enthusiastic, arrogant, skeptical, and stubborn, but also a gallant charmer with great charisma. Lepsius came to Paris in 1833, one year after the death of Jean-François Champollion, who had cracked the hieroglyphic code in 1822. Although Lepsius was unable to read the hieroglyphs, he was fascinated by Champollion's work, and he realized, intuitively, that the decoding was not complete. Lepsius initiated a correspondence with Ippolito Rossellini, one of Champollion's students. Three years later, they met in Pisa. By that time, Lepsius had learned to read the hieroglyphs. He quickly realized that Champollion understood the hieroglyphics as abbreviated words which conveyed meaning but still remained incomplete. Lepsius added one very significant discovery to Champollion's translations: The hieroglyphs were not only abbreviated words but also symbols for sounds and syllables. With determination and tenacity, Lepsius copied and translated virtually every hieroglyphic text available in Europe.

In 1841, several of Lepsius's friends—Alexander von Humboldt among them—appealed to his majesty King Friedrich Wilhelm IV of Prussia, in his wisdom and generosity, to fund an expedition to Egypt. The leader of the expedition was to be Richard Lepsius, who, by then, had published several works about Egypt. The king gave his consent. In August of 1842, the Royal Prussian Egypt Expedition set sail from Hamburg. The team included a painter, a draftsman, a stucco expert, and two architects—plus thirty crates of materials. The Prussians were not pinching pennies.

After arriving in Egypt, Lepsius met with the viceroy, who provided him with several letters of authorization and asked him, expressly, to present as gifts to the Prussian king any archaeological artifacts which Lepsius deemed worthy. No one had begun to catalog any archaeological findings in Egypt, no one had heard of Auguste Mariette, and the Europeans did as they pleased in Egypt. Thus, Lepsius sent two hundred crates full of archaeological treasures to Berlin over the years (some of which Egypt would now like to see returned). Richard Lepsius was not an overly sensitive man. On the king's birthday, he had the Prussian flag hoisted on the Great Pyramid at Giza to the tune of the Prussian national anthem. At Christmas 1842, Lepsius ordered his Egyptian workers to haul wood to the tops of the three great pyramids so that they could light bonfires to the accompaniment of "Silent Night." In a humorous account of the greatest archaeological adventures, Philipp Vandenberg describes Richard Lepsius, commissioned by Friedrich Wilhelm to lead this expedition, spreading Christmas cheer dressed in a dark suit and holding a candle in his hand. The sarcophagus of King Khufu was adorned with a young palm tree decorated with burning candles.[12]

Lepsius could be quite sentimental—and he knew how to sing! All of Cairo was amazed.

What Did the Royal Prussian Egypt Expedition Find?

In May 1843, the Royal Prussian Egypt Expedition departed from Giza with a new destination in mind: the Labyrinth. Lepsius was familiar with the reports of Herodotus, Strabo, and others, and he knew exactly where to locate the maze. How?

One hundred and twenty kilometers southwest of Cairo, the desert blooms in a small area of fertile land: the oasis Fayyum. The oasis, with its lush vegetation, has been connected to the Nile by means of a canal, the Bahr Yusif, for thousands of years. Twenty-five kilometers northwest of the town of Fayyum, one encounters the flat Lake Qarun, which many archaeologists believe to be Lake Moeris. Thirty-seven hundred years ago, the pharoah Senwosret II (1897–1878 B.C.) had a pyramid built in his honor out here in this heavenly oasis, so starkly surrounded by cracked rocks and sand dunes.

According to Diodorus Siculus, the builder of the Labyrinth was named Mendes, also called Marrhos.[13] Manetho calls the same ruler "Lamares," and Plinius attributes the name of the lake to the person: "Moeris." "Marrhos" is also the throne name of Amenemhet III (1844–1797 B.C.), and this pharaoh moved his summer residence, including his pyramid, to Hawara, forty kilometers from the shores of the present-day Lake Qarun. Moreover, the ancient capital of the oasis Fayyum was named "Krokodeilon polis"—city of crocodiles. It once served as the central place of worship for the crocodile god Sobek. It was easy to make the connection: The Labyrinth was supposedly located near the city of crocodiles;

the builder of the maze was "Marrhos" and this happened to be the throne name for Amenemhet III. This pharaoh had a pyramid constructed in his honor at the oasis Fayyum. Consequently, the Labyrinth had to be located near this pyramid. Or didn't it?

Lepsius was not the first to search for the maze near Fayyum. In 1714, the French explorer Paul Lucas had sojourned on the shores of Lake Qarun because he assumed that he would find there the ruins of the two pyramids which protruded from the water in Herodotus' account. However, after inspecting the shoddy, waterlogged boats in which the fishermen planned to take him across the lake, Lucas abandoned his endeavor.

In January of 1801, Dr. P. D. Martin, an engineer with Napoleon Bonaparte's Egypt forces, rode all the way across the desert to visit the oasis Fayyum. The bedouins were impressed by his willingness to endure such physical hardships, and they answered his questions readily. Still he did not find the Labyrinth.

In 1828, the king of France, Charles Philippe X (1757–1836), appointed the studious translator of hieroglyphs Jean-François Champollion to lead an expedition to Egypt. Champollion, a sensitive and highly intelligent man, searched in vain for the Labyrinth at the oasis Fayyum.

Finally, one year prior to Lepsius's visit, a French expedition reached the pyramid of Amenemhet III. They found some small walls and broken pillars, but nothing reminiscent of a gigantic maze of buildings.

After Julius Caesar won the battle of Antioch on August 2, 46 B.C., he sent three immortal words to Rome: "Veni, vidi, vici!" I came, I saw, I conquered! The same thing happened with Richard Lepsius. He came, he saw, and he conquered. Completely sure of himself, he noted shortly after his arrival that they moved camp on May 19, 1843 and pitched their tents in Fayyum on May 23, on the ruins of the Labyrinth. He added that the location of the maze had long been correctly identified and that there was no doubt in his mind he'd rediscovered it the moment he first laid eyes on it.[14]

The letters Lepsius sent to Berlin are further evidence of his premature conviction. He stated that his crew had been camped on the south side of the pyramid of Moeris since May 23, on the ruins of the Labyrinth. He maintained that he was fully entitled to make those assertions and that he'd immediately realized the nature of his discovery.

These statements by Richard Lepsius took care of Herodotus' Labyrinth, as far as the scholars were concerned. It was conveniently cataloged and labeled, even though a closer look would have revealed that nothing agreed with the descriptions provided by the ancient historians. Lepsius hired men and children from the surrounding villages. They were supervised by overseers and received bread. They were counted every morning and paid every evening. Every man was paid one piaster, every child half of one piaster, sometimes thirty para if they were particularly diligent. The men were required to bring a rake and a woven basket; for the children, a small basket was sufficient. Lepsius had the workers dig trenches in five different locations simultaneously. The men filled the baskets; the children and the old men hauled off the rubble. The procession of the baskets was supervised by the overseers, who also encouraged the busy laborers to sing during their work.

After a few short days, Lepsius had cleared an area which revealed the remains of granite and limestone columns, shimmering "almost like marble." Of course, Herodotus had described real marble stones, "exquisitely fitted together." Lepsius was thrilled to find "hundreds of chambers, next to and on top of each other, some small, even tiny, others large or even huge, supported by columns, with no regular pattern for exits and entrances, so that the descriptions by Herodotus and Strabo are totally justified in this regard."

Are they really?

Archaeologist vs. Historian

Where are those walls "carved all over with figures" (Herodotus)? Where are the convoluted, winding paths (Strabo)? Where is the ceiling made from a single stone? Where are the covered passages "roofed with single slabs of surpassing size" (Strabo)? Lepsius dug up small, even tiny chambers, while Herodotus walked "from chambers into colonnades, and from the colonnades into fresh houses." Neither Herodotus nor any of his successors made any mention of having had to crawl or stoop through the chambers.

Regarding the entire structure, Lepsius noted that the layout involves "three large building complexes, each three hundred feet wide, enclosing a rectangular courtyard, six hundred feet long and close to five hundred feet wide. The fourth side, one of the shorter ones, is bordered by the pyramid behind it, which is three hundred feet wide and therefore does not touch the sides of the other buildings."[15]

How does this coincide with the twelve roofed courts which Herodotus describes? With the "large figures engraved on it"? With the chambers exceeding "all other human productions"? Lepsius personally attests to not having found any inscriptions in the ruins of the large chambers. Herodotus, on the other hand, marvels at the walls "carved all over with figures." Lepsius claims that a long wall divides the central square into two halves, while Herodotus speaks of a single wall that "surrounds the entire building." Two thousand years ago, Plinius reported that ninety steps descended to the underground galleries. Ninety steps? That is quite deep. If we estimate each step at a moderate height of twenty centimeters, we would arrive at a depth of about eighteen meters below the surface level. Lepsius saw no trace of any such stairs. According to Plinius, there were several houses in underground vaults. Lepsius never entered any subterranean houses. The Royal Prussian Egypt Expedition never discovered sepulchers or sarcophaguses of ancient mythological pharaohs.

Sic transit gloria mundi. So passes away the glory of the world.

The preconceived notion that the pyramid of Amenemhet III was the one under which the Labyrinth would be found was wrong from the very start. If Lepsius had kept a clear hear and had investigated further, he probably would have realized this. It is true that the name "Marrhos" which Diodorus refers to is also the throne name for Amenemhet III, but why did Lepsius refuse to look beyond that? After all, the ancient historians listed many other names besides "Marrhos" in their descriptions of the Labyrinth.

Here is a list of names:

HERODOTUS: twelve kings, among them by name Psamtik, who "ruled Egypt for fifty-four years."

DIODORUS SICULUS: Mendes or Marrhos, as well as Psamtik of Sais and Moeris.

PLINIUS: Petesuchus or Thitoes, as well as Motherudes and Moeris.

MANETHO: Lamares.

There is no reason to believe that, in our search for "Marrhos," Amenemhet III would be a more valid option than any of the others. In fact, the evidence speaks against his being the builder of the Labyrinth. And the evidence is very unequivocal.

Contradictions Galore

Do you recall our eyewitness Herodotus? "At the corner of the Labyrinth stands a pyramid, forty fathoms high, with large figures engraved on it. . . ." And Strabo agrees: "At the end . . . is . . . a quadrangular pyramid, which has sides about four plethra in width. . . ."

According to Herodotus, the pyramid is supposed to be 71 meters wide; according to Strabo, the width is 120 meters. The pyramid of Amenemhet III in Hawara, on the other hand, measures 106 meters. Nothing matches. Herodotus and Strabo agree that the pyramid is located at one of the corners of the Labyrinth. This is not the case in Hawara. The pyramid of Amenemhet III is not in a corner, but rather on the same axis as the temple ruins. Herodotus, the eyewitness, saw large figures engraved on the walls. You will not find large engraved figures in Hawara because the pyramid is built from mud bricks. It is impossible to engrave any kind of figure on dried mud bricks, let alone "large" ones.

Just imagine the perversity of it all: All our historical eyewitnesses describe the Labyrinth as a wondrous structure with walls that are "carved all over with figures," an edifice that excels "all other human productions" (Herodotus) and cannot be surpassed (Diodorus), made from gigantic slabs of stone of "surpassing size" (Strabo) with "large blocks of syenite" (Plinius). (Syenite is a limestone resembling marble.) And now imagine that the same king Amenemhet III, who dazzled the world with this superb structure, had a burial pyramid built for himself with the cheapest, flimsiest, and most fragile building material possible: mud bricks! That's like a square peg in a round hole! The facts, clearly, speak for themselves.

Each and every pharaoh was proud of his accomplishments. The rulers along the Nile immortalized their feats on tablets and inscriptions so that later generations would know which temples they built or restored. If Amenemhet III had been the builder of the maze (which "surpasses the pyramids," according to Herodotus), inscriptions would abound applauding this brilliant achievement and heaping praise on the pharaoh responsible for it. No such inscriptions have been found. In one of the chambers and on some column fragments, Lepsius encountered nameplates with "Amenemhet III" on them. He concluded, quite accurately, that this proved the identity of the builder and owner of the pyramid. Forty-five years after Lepsius, the British archaeologist

Sir Flinders Petrie (1853–1942) discovered the undisturbed sarcophaguses of Amenemhet III and his daughter inside the pyramid. The sepulcher consisted of a single block of yellow quartzite lowered into the ground. The burial vault was covered by three heavy quartzite slabs, 1.22 meters thick,[16] enough to bear the weight of the mud bricks above. Finally, canal workers discovered a limestone statue of Amenemhet III, about 1.6 meters high, near the pyramid.

There is not the tiniest trace of any hieroglyph, on any of these artifacts, that would indicate that Amenemhet III was the builder of the Labyrinth. Flinders Petrie found the sepulcher of Amenemhet III completely intact. It is inconceivable that the pharaoh would not have been extolled in his tomb for his accomplishment if he had indeed been the brilliant mind behind the construction of the Labyrinth. No pharaoh would have passed up such a grand opportunity to attain immortal glory!

It is curious that the same pharaoh Amenemhet III also had a pyramid built in Dashur, twenty kilometers south of Cairo. It is popularly known as the "black pyramid" because it is constructed of dark gray Nile mud bricks. The very top stone of the pyramid, the so-called pyramidion, is 1.4 meters high and made of black granite. It is currently exhibited at the Egyptian Antiquities Museum in Cairo. The god Horus spreads his wings over the hieroglyphic inscriptions, which clearly identify Amenemhet III as the builder of the pyramid. Not a word is said, however, about Amenemhet III being the creator of the phenomenal Labyrinth. A few years ago, archaeologists from the German Archaeological Institute in Cairo uncovered next to the previously known empty sarcophagus made of rose granite two additional sarcophaguses belonging to two of Amenemhet's wives. Again, there is no indication that their lord and master was the builder of the incomparable Labyrinth.

I have been reading books about Egypt for many years, books written by intelligent and astute archaeologists. All of them maintain that the Labyrinth is located below the pyramid of Amenemhet III in Hawara. Everybody mentions Lepsius and his discovery of the maze. These scholars seem to be climbing a spiral staircase without ever looking back; they simply accept the nonsense that the person in front of them is handing down to them. And we know now that many of the small walls and chambers which Lepsius' singing workers uncovered date back to Greek and Roman times! Amenemhet III was responsible for the construction of the mud pyramid and some nearby temples—but these buildings are as unrelated to Herodotus' Labyrinth as Beethoven's Fifth is to music from the Hit Parade.

It is evident that the archaeological remains do not coincide with the descriptions provided by ancient historians, but wait until you consider the geographical location: The situation gets positively grotesque.

A Lake Evaporates

Herodotus assures us that the Labyrinth and the pyramid are adjacent to Lake Moeris. He describes this lake as "astonishing." Its "circumference is sixty schoenes, or three thousand six hundred furlongs, which is equal to the entire length of Egypt along the sea-coast." In comparison: Herodotus' lake would have had a circumference of 640 kilometers. Lake Constance, one of the largest lakes in Europe, has a circumference of 259 kilometers: 160 kilometers of shoreline in Germany, 72 kilometers in Switzerland, and 27 kilometers in Austria, with a total area of 538.5 square kilometers. Consequently, Lake Moeris, as described by Herodotus, would have had twice the circumference of Lake Constance.

It is possible that the Egyptians supplied Herodotus with exaggerated figures or that the figures were translated wrongly from Egyptian into Greek. The figures notwithstanding, it is certain that the Labyrinth and the pyramid were located on the shores of the lake, because Strabo also emphasizes the size of Lake Moeris and the fact that the pyramid is set along its shore. Strabo insists that he visited the site personally: "Our host, one of the officials . . . went with us to the lake. . . ." Subsequently, in the presence of Strabo and the host, a priest was trying to feed a sacred crocodile, which was dozing on the beach, too lazy to eat the bread being offered.

Diodorus Siculus also pays tribute to Lake Moeris.[17]

> Moeris succeeded to the sovereignty twelve generations after Egyptus. In Memphis he built the north propylaea. . . . Upriver from the city, at a distance of ten schoinoi, he excavated a lake which is admirable for its usefulness and unbelievable in the scale of the labor involved. For they say its circumference measures three thousand and three hundred stades and its depth, in most places, fifty fathoms. And who indeed, contemplating the immensity of this endeavor, can help but wonder (and with good reason) how many myriads of men and how many years it took to finish the work?

Later in his book, Diodorus supports Herodotus' contention that the water level in the lake was regulated by large locks, which were opened or closed depending on the water level of the Nile.

The Labyrinth, pyramid, and lake belong together. Geologists tell us, however, that there has never been a lake near the Hawara pyramid.[18] They base their conclusions on the sedimentation of the soil. Moreover, two additional reasons speak against the existence of such a lake. The pyramid of Amenemhet III is made of thousands of mud bricks. The combination of mud and water, in this case, would be lethal to the pyramid; its foundation would be weakened. The rooms and chambers dug up by Lepsius's expedition would have been flooded unless they had been strengthened to withstand the lake water. But no waterproof protective walls were found in Hawara.

Today, a flat lake, Lake Qarun, is situated about twenty-five kilometers northwest of the town of Fayyum. However, it cannot be identical with Lake Moeris as the ancient historians described it. First of all, it is forty kilometers from the Hawara pyramid; second, it is a natural lake not fed by artificial channels; third, Lake Qarun is bordered by sizzling desert sands on three sides and some vegetation and a few tourist hotels on the fourth; and most inconveniently, it is below sea level. Richard Lepsius was well aware of this. He wrote that the lake does indeed rise if the Nile carries a lot of water and therefore increases its input to Qarun, but he realized that the lake is too low to ever let any of its water flow back. The whole province would have been flooded before the water would flow back to the valley. During Lepsius's visit, the water level of Lake Qarun was about twenty-one meters below the point where the canal issues into the lake, and Lepsius was convinced that it had never risen significantly above that. He found proof of this in the ancient temple ruins along the shores of the lake. Lepsius concluded that Lake Qarun definitely was not the lake that was adjacent to the Labyrinth or the town of Arsinoë, now Medinat Fayyum.[19]

In spite of this fact, Lepsius insisted on having identified the location of the Labyrinth. Together with three of his colleagues, he had visited the remains of a dam which he interpreted as an artificial mound created for Lake Moeris. Lepsius even studied the ruins of two buildings that he initially believed to be the two pyramids which, according to Herodotus, protruded from the lake. But after a short excavation, Lepsius concluded that it was certain they had not been sitting in a lake.

Even people like Lepsius who are fervently convinced that the Labyrinth is located in Hawara should have been tipped off by the distances specified in the works of Herodotus and others. Diodorus Siculus, for instance, wrote that King Moeris had the artificial lake constructed "ten schoenes above the city of Memphis," which would place it roughly at the same latitude as Dahshur, about seventy kilometers northeast of Hawara, as the crow flies. Strabo described the lake as a huge body of water, with beaches, comparable to the ocean. Herodotus specified that "none of the land below Lake Moeris [was] then showing itself above the surface of the water. This is a distance of seven days' sail from the sea up the river."

And finally, Herodotus gives us the following clue: "The natives told me that there was a subterranean passage from this lake to the Libyan Syrtis, running westward into the interior by the hills above Memphis."

A Personal Inspection

The taxi driver smiled when he saw me emerge, camera-laden, from my hotel. He knew me well, because I had employed his services on several previous days. This way I avoided the hassle of having to choose from among the hordes of competing drivers camped out in front of every hotel in Cairo, and I also saved myself the trouble of having to negotiate for a daily rate each time. The taxi driver knew what he could expect, and so did I. Moreover, his vehicle, an older, black American car, was in surprisingly good condition, a fact that is crucial in Egypt, where roads rather quickly turn into desert or gravel paths. The driver's name was Kamal, and he had studied Egyptology at the university in Cairo for four years. He decided to become a cabbie because he could make more money driving a taxi than by holding a regular job in his field. He knew some English and was adept at keeping the pesky souvenir dealers at a distance. This is particularly refreshing if one considers that most of the guides and dealers know each other and that the driver usually shares in the dealer's profits.

We drove along the main highway toward Giza, past caravans of honking, fume-billowing vehicles, past the great pyramids, continuing southwest into the desert. The road to Fayyum, about 106 kilometers of straight asphalt, is flanked on both sides by rusty cars and gutted buses and trucks gradually disappearing into the desert sands.

"What are you looking for out there?" asked Kamal.

"I want to see the pyramid of Amenemhet III at Hawara."

"It's not worth it," said a laughing Kamal, expertly. "Nothing much to see, just dried bricks."

"I know. I want to see it anyway."

Kamal laughed again. "You people are all a little strange, if I may say so. No Egyptian would visit the Hawara pyramid on his own free will."

A brief thought about Herodotus flashed through my head. Herodotus would have been dependent on camels to get from Giza to Hawara—two days' journey. Certainly, he would not have forgotten to mention such a journey in his report. Today, we are fortunate enough to reach the outskirts of the oasis in two hours, thanks to our motorized "camels." (Actually, it is misleading to use the term "oasis" to describe Fayyum, because the "oasis" is totally dependent on the Nile for its water, via the Bahr Yusif Canal. Fayyum, with its four thousand square kilometers, does not produce any water independently. I will continue to use the term "oasis," however, because any water in the middle of the desert is an oasis to us, no matter what its origin.)

The fertile area of Fayyum, surrounded by desert, is irrigated by 324 canals with a total length of 1,298 kilometers. According to official reports, this system is supplemented by 222 water channels with an additional 964 kilometers.[20]

The car radio was breathing a prayer into the air; a muezzin was reciting the prayer, while the crowd hummed. Kamal, while driving, bowed three times over his steering wheel. "It's about the water," he explained later. Sheik el-Azhar, the leader of the Sunnites, had called his followers to ask Allah for water. The summer of 1988 marked the seventh year of a drought in Ethiopia. Without rain, there would be no water in the Nile; without water from the Nile there would be no water in the muddy canals, and without that water there would be no agriculture.

"Egypt has the Nasser Dam near Aswan," I said sympathetically. "Doesn't it regulate the flow of the Nile?"

Kamal smiled again. He smiled all the time. This time he was smiling about my ignorance.

"The water level of the dam has decreased by twenty-five meters in the last few years. If it doesn't rain in Sudan or Ethiopia within the next couple of months, the turbines will have to be shut off. That would be bad for everything that relies on electricity! If that happens, the little water that's left in the Nile will not be sufficient to feed the thousands of canals on both sides of the river. The fields will dry up. Do you know what that would mean for fifty-three million Egyptians?"

I could imagine it. Throughout its history, the whole country has been dependent on a single source of water. Currently, about 2.6 million hectares of land are irrigated, guzzling up approximately 49.5 billion cubic meters of water annually. An additional 3.5 billion cubic meters are required for drinking water. Whoever Pharaoh X was who commissioned the construction of Lake Moeris, he must have been a wise and provident ruler.

After ninety kilometers, we encountered the first vegetation along the road. Merchants waved from their stands, offering rose bouquets, garlands of onions, and live turkeys. After we turned a corner, we saw the first canal. Children were splashing happily in the sluggish, brown water. I told Kamal to stop. He was not smiling now; his face had become sullen.

"Bilharziasis?" I asked. Kamal nodded.

The canals were polluted with bilharzia parasites, microscopic flatworms named after the German physician Theodor Bilharz (1825–1862), who discovered them. The parasites enter the system through the skin and lodge in the liver, where they multiply, causing diseases of the liver, intestines, and genitals. These conditions used to inevitably be fatal; today, we have medications to fight bilharziasis. The Egyptian government, in cooperation with the World Health Organization, has been battling this insidious disease for many years. The parasites multiply rapidly in the shallow mud along the canals, where the water is almost stagnant.

"Why are the children allowed to swim in the water?"

Kamal shook his head. "There are many information campaigns on TV, on the radio, in the schools, even in comic books, but many of the farmers refuse to see the danger. They have faith in Allah."

They are God-fearing, honest, and humble, these diligent farmers and their families, who work in the sprawling, hot fields all day, year in and year out. They plant cotton, or, depending on the season, beans, corn, rice, cucumbers, potatoes, onions, garlic, cauliflower, and watermelon. Harvesting machines are rare. The bent backs of women and children are more economical. Stands of palm trees afford some shade. Every part of the tree is used, from the trunk to the individual fibers. Some women weave baskets in front of the clay huts; others make clay bowls, lamps, and figures. Children decorate the products in bright colors. Time seems to stand still.

Kamal pointed ahead. "That's Medina, the capital of

the oasis. Nowadays, it's mainly called 'the Fayyum.' Supposedly, it used to be called City of Crocodiles."

" 'Supposedly'?"

Kamal squirmed and smiled, the thought of the children swimming in the canals and the bilharziasis temporarily wiped from his mind. "We know for sure that it was called City of Crocodiles at one time. But if we talk about the City of Crocodiles, everybody thinks of the place that the ancient writers described: the City of Crocodiles on Lake Moeris."

"And? Aren't they one and the same?"

My archaeologically well-versed cabdriver shrugged his shoulders and smiled impishly. "There were a number of crocodile cities in ancient Egypt, and every larger temple between the Nile delta and Aswan worshiped the crocodile in some form or other. Here in Fayyum, virtually every village had a crocodile shrine. It is difficult to say which city of crocodiles Herodotus was referring to."

I thought to myself that this did not resolve any of my questions.

Slowly, we zigzagged between the throngs of people, past a truck loaded with camels. (The camels are lazier too these days!) Kamal stopped the car.

"Since you are here, you should take a look at this," he said.

In one of the canals crossing the center of town, four huge, dark brown water wheels were churning through the water. They were groaning, cracking, squeaking, and moaning, as if inside of them hidden ghosts were being urged on by invisible whips. The wheels turn continuously. I learned that the oasis Fayyum has approximately two hundred such water wheels. The wheels direct the water into the different canals and elevate the water level, all without electricity or any other outside source of energy. The only source of energy is the flow of the water. The wheels display superior craftsmanship. They have wide, simple paddles attached to them, which dive into the water with each turn. Powered by the water current, the wheel is forced to rotate. The sides of the wheels are equipped with scoops, which fill themselves with water as they pass through the current and then release the water into the higher canal. The maximum height of this perpetual water lift depends on the circumference of the wheel. It is amazing what ingenious ideas and inventions sprang forth from human imagination thousands of years ago!

About ten kilometers southeast, near the small village of Hawara, the dark gray pyramid of Amenemhet III rises toward the sky. From afar, it reminded me of a cup of pudding turned upside down, with a flattened top and numerous dents. I came without prejudice, and I probably would have jumped for joy if I had found even a trace of the Labyrinth anywhere near the pyramid. In a makeshift shelter, a lonely policeman was startled out of his sleep by our arrival, and next to the shelter, a sign with black borders proclaimed: "Labyrinth. 305 x 244 m/3000 Rooms." There was no trace of "3000 Rooms" anywhere.

I poked around in the sand for a few hours, climbed over some low walls dating back to the times of the Ptolemies and the Romans, shone my powerful flashlight into holes and shafts, and searched eagerly for walls "carved all over with figures" (Herodotus). The only thing that reminded me even faintly of ancient temples were a handful of blocks of reddish Aswan granite. There were no remains of "slabs of surpassing size" (Strabo) or "large blocks of syenite" (Plinius), let alone the skeleton of a structure that exceeded "all other human productions" (Herodotus).

Step by step I climbed up the southwest corner of the pyramid, searching for something unusual under the grayish-black mud bricks—perhaps a granite block that, in the times of Herodotus, could have supported the weight of the "large figures" engraved in the pyramid (Herodotus). I found nothing! The pyramid is partially destroyed. Local builders have used the prefabricated bricks for their houses. The top of the pyramid is missing completely; one could pitch a tent up there. Originally, the pyramid was covered with a layer of limestone, but there is not a trace of it left. Rain has eaten gutters into this mountain of mud bricks; many of the bricks, originally about fifty centimeters long, have been eroded or worn away over time. The raw material for the bricks was pressed between boards and dried in the sun. As a result, the bricks are porous, blemished with blades of straw, dried grass, and small rocks.

Not even an Egyptian Michelangelo could have engraved "large figures" into these bricks; the mud would never have been able to support the weight of such sculptures. Critics will argue that the figures have long fallen from the walls of the pyramid, that the walls "carved all over with figures" have crumbled over time. But why should that be the case only here, at the Hawara pyramid and the alleged Labyrinth? Broken statues of many pharaohs have been found at other sites, and at the wonderful Egyptian temples which attract millions of tourists each year the walls "carved all over with figures" have not disappeared into thin air. We know that they still existed in the Labyrinth at the time of Herodotus' visit.

But never mind: At least the remains of the huge stone slabs "of surpassing size" should still be lying around. Nothing! There simply is no trace of anything!

Nor does the view from the top of the pyramid, fifty-eight meters high, produce better results. All the eye can see are some low walls, sand dunes, and, farther off, electrical towers, with a canal diagonally dissecting the whole area and in the distance some cultivated fields.

These pitiful piles of rubble are supposed to be the ruins of the exquisitely praised Labyrinth?

Kamal had discovered a skull amidst the rubble. The policeman placed it on a wall. I stared into the empty eyes of the skull and wondered, briefly, if the deceased may have met Strabo or Herodotus a long time ago. If the dead could speak . . . I would ask the skull where to look for the remains of the extraordinary Labyrinth. Kamal was laughing out loud. I felt as if the skull were laughing, too, in chorus with all the gods of ancient Egypt.

The Labyrinth a Pile of Rubble?

In 1888, forty-five years after Richard Lepsius, the British archaeologist Sir Flinders Petrie came to this site. He concluded that the chambers which Lepsius had unearthed were merely the ruins of the Roman village in which the destroyers of the Labyrinth had lived.[21] The maze itself, according to Sir Petrie, had been completely dismantled, the only remaining evidence of its former existence being a hollow in the ground filled with countless small fragments. Petrie asserted that it was extremely difficult to put something back together from such a dearth of fragments, yet he proceeded to attempt just that.

I wish he had left it alone! His own version of the Labyrinth, a plan with countless chambers and columns, is as incongruent with the descriptions of the ancient historians as the plan submitted by Lepsius. In Petrie's plan, the temples and colonnades run in straight, parallel rows. Strabo, however, described winding, convoluted paths and warned that it was impossible to find the exit without a guide. Plinius, too, emphasized the confusing, crisscrossing passages. I fail to see why visitors to the kind of labyrinth that Sir Flinders Petrie drew would have had any problem whatsoever finding the exits; they are all aligned with the exactness of parading soldiers. Petrie's plan included several detached temples facing each other at great distances. On the other hand, Herodotus, the eyewitness, spoke of roofed courts that were very close together. Petrie uncovered the remains of walls along the southern and western borders of the site. Herodotus, however, saw "a single wall" that surrounded the entire building. Petrie's wall cannot be identical to the circular wall described by Herodotus, because if it were, traces of it would have been visible in the northern and eastern sections as well.

Petrie's plan of the Labyrinth is confusing and full of contradictions. At first, the building is square, then it is rectangular, and finally it is round. Evidently, Petrie, like his predecesor Lepsius, tried to fit the few available fragments into a preconceived plan. With this method, any archaeological pile of rubble can be transformed into a labyrinth. Petrie's interpretation ultimately failed because he was unable to magically transport the huge Lake Moeris to this site in Hawara, and in spite of fervent excavations, he never found fifteen hundred subterranean chambers.

H. L. Mencken once said that for every problem there is a solution that is simple, clear, and wrong.

Where is the Egyptian Labyrinth? Did Herodotus and his successors lie to us? Did this building which, according to Herodotus, surpassed description never even exist? Or did the term "labyrinth" mean something different to the ancient historians than it does to us? Are Herodotus and his colleagues mere plagiarists who stole their sensational stories from other sources?

Labyrinthian Complications

Today, as in ancient times, a labyrinth denotes a maze, a system of caves with convoluted passageways or a building with an intricate internal system of stairways, winding corridors, and rooms. Myths about labyrinths originate in the Stone Age.

Engraved pictures of labyrinths have been found on rocks and cave walls in northern Africa and southern France, on the islands of Crete and Malta, and in southern India, England, Scotland, and the United States. The maze motif had international appeal even in prehistoric times. The later labyrinthian decorations of Greek geometric vase paintings and Mexican and Peruvian pottery are surprisingly similar as well.[22] We may wonder about the reasons behind this global pattern. What prompted Indians in Arizona to scratch labyrinthian pictures into the rocks, if they had no contact with their prehistoric European colleagues? Did the Stone Age people of all continents examine the open skulls of their enemies and use the winding passageways of the human brain as their original model? Were they playing hide-and-seek or "What am I thinking?" Were they trying to get a physical grip on the thoughts of the gray cell matter? I suspect that the thought processes of these Stone Age people were much less labyrinthian than ours.

Scientists searching for certain motifs can be likened to the shipwrecked Robinson Crusoe, who, one day, discovers a footprint in the sand. The trail leads into uncertainty. The labyrinth is like a monster with a thousand tentacles: We can never quite grasp it and always retreat from it because of our fear of the unknown. Greek legend has it that Daedalus, an Athenian architect and inventor, constructed a maze in Knossos on the island of Crete. Originally, the complex, with its convoluted passages from which no one could escape without help, was intended for the Minotaur, a man-bull hybrid. Both Diodorus Siculus and Gaius Plinius Secundus affirm that the labyrinth on Crete was merely a reduced copy of the original Labyrinth in Egypt.

Sir Arthur Evans, the great excavator of Crete, found no trace of a labyrinth. This led archaeologists to believe that the term "labyrinth" originally was not intended to describe a single building, but rather an entire city with a network of streets. Jan Pieper, who has studied the labyrinth myth, asserts that there is indeed reason to believe that historically the myth is based not on a single gigantic building with a labyrinthian structure, but rather on a city full of people, which to the pastoral tribes must have inevitably appeared labyrinthian, so that they could hardly expect anything but a bull-headed, man-eating monster at its center.[23]

Although this argument is alluring because of its simplicity, it will not unlock the gates of the labyrinth. After all, the Stone Age artists on the various continents had no labyrinthian cities teeming with people as models to base their creations on.

"All of us err, but each of us errs differently" (Georg Christoph Lichtenberg, 1742–1799).

Ancient Liars?

In Egypt everybody certainly errs differently, because eyewitnesses assure us that they have personally entered the Labyrinth. Herodotus emphasizes four times on one page that he is speaking from personal experience. Why should the "Father of History" lie to us in quadruplicate in the space of a single page? After all, he was telling the truth in other places. Why should Strabo reaffirm Herodotus' lies 423 years later, and even add some of his own in the bargain? That would mean the little anecdote about his host, "one of the officials," and the priest who was feeding a crocodile on Lake Moeris is pure invention. And what about Gaius Plinius Secundus, who wrote that he was surprised at the stones he saw at the entrance to the maze? Was he surprised on paper only? Why would he want us to believe that he was tired from walking through the impossibly convoluted passages if he never actually walked a single step and therefore could not have become tired? How could he descend ninety steps into galleries that never existed?

I believe the old writers. The Labyrinth which "surpasses the pyramids," was "a little above Lake Moeris" (Herodotus). Can a lake with a circumference of six hundred and forty kilometers disappear into thin air? I mentioned earlier that Herodotus' measurements may have been exaggerated, but even a lake of that size can evaporate rather quickly. The Nasser Dam near Aswan has a length of over five hundred kilometers. Seven years of drought in Sudan and Ethiopia were sufficient to lower the water level by twenty-five meters. Even droughts lasting more than seven years are no reason to expect the end of the world. The Old Testament tells of the seven years of drought in Egypt which the people were able to survive because of Joseph's foresight.

In Herodotus' account, Lake Moeris was fed by a canal from the Nile. If the river ran dry, the canal was bound to dry up as well. Most likely, the locks to Lake Moeris were closed during any extended drought so that the crucial water supply along the Nile could be maintained. Such water shortages were not uncommon in the land of the pharaohs. Lake Moeris supposedly fed some of its water back to the Nile. But suddenly things were different.

Since Lake Moeris existed during the lifetime of Herodotus and since Strabo was able to feed a crocodile on its shores 423 years later, the lake must have gradually dried up during the Roman-Christian era. The powerful empire of the pharaohs had been subdued. No insightful ruler gave the order to restore Lake Moeris, to dig up the canals, and to repair the locks and dams. In *Geography*, Strabo describes several large canals and smaller lakes in

Egypt, which were navigable and supplied the sprawling communities with water. What is left of these?

A few years of drought combined with a few years of lethargy allowed Lake Moeris to evaporate. Diodorus Siculus hypothetically asked how many people and how many years had been required to create Lake Moeris. Now that the lake was beginning to dry up, the manpower was lacking, as was the command structure that could have motivated and organized a new army of workers. It was the beginning of the end, not just for Lake Moeris and the Labyrinth, but for Egypt in general. Temple cities which had been carefully maintained for thousands of years were being abandoned, giant pyramids and the huge sphinx of Giza were being devoured by the desert sand—our present day excavations are proof of this.

But sand not only swallows anything and everything—it also preserves. The Labyrinth praised by Herodotus, with its ornately carved walls, fifteen hundred subterranean chambers, and priceless tombs of twelve mythical pharaohs, is waiting for a modern-day Heinrich Schliemann. Chances that the Labyrinth will be found are not too bad, thanks to the numerous clues the ancient historians left for us. If we summarize the information contained in the accounts of Herodotus and his colleagues, we must conclude that the Labyrinth is located seven days' journey up the Nile, on the Libyan side, a little above the city of Memphis, at the entrance of the canal into Lake Moeris. The lake extended in a north-south direction and was situated in the Arsinoë region. Last, but not least, the canal feeding the lake was connected to the Nile and was regulated by locks and dams.

Very simple, isn't it?

The Last Chance

It is a simple recipe: Start with a small airplane, maybe a helicopter, and circle above the pinpointed area early in the morning or in the evening.

You may have to stir for a while before the dough is ready. Maybe you will have to travel up and down the Nile for a month before you will find a trace. A trace of what? A trace of the canal, my son! What canal? The canal which flowed by the Labyrinth, my son! But the canal doesn't exist anymore. Precisely, my son!

Aerial photography is a powerful archaeological tool. Canals that have been dry for thousands of years may still be partially visible from the air. Somewhere above Memphis, a canal must have issued westward from the Nile. Its route should still be traceable. If there is no such canal, the only other option is the old Bahr Yusif canal, on whose banks vegetation continues to thrive. If a canal is discovered from the air, we can follow its course to the end. That would be the location of Lake Moeris, where the Labyrinth is waiting to be unearthed. If, however, the Bahr Yusif is the only possibility, we should be able to locate the remains of ancient locks along its original course. They would lead us to the Labyrinth, for the ancient historians agreed that the maze was situated at the entrance of the canal.

The connection is easy to make, although drawing connections can be confusing. Mark Twain once said that although there is probably a connection between a rose and a hippopotamus, no young man would think of handing his lover a bouquet of hippos.

✧ 3 ✧

The Nameless World Wonder

"Man fears time—Time fears the pyramids."

EGYPTIAN PROVERB

"Pickles Cause Airplane Crashes, Car Accidents, War, and Cancer." This amazing headline shocked the academic world in the summer of 1982 when it appeared in the *Journal for Irreproducible Results*. The statistical evidence was solid. After all, 99.9 percent of all cancer victims had, at some point in their lives, eaten pickles, as had virtually all soldiers; 99.7 percent of pilots and car drivers also admitted to consuming pickles occasionally. Naturally, the report was a hoax, because the *Journal for Irreproducible Results*, published quarterly in Park Forest, Illinois, specializes in satirizing scientific studies. It demonstrates how statistics, misleading hypotheses, and erroneous interpretations can be used to buttress any claim.

Let us experiment with a similarly strange hypothesis connecting the Egyptians' consumption of onions to the construction of the pyramid. At the time the Great Pyramid of Giza was erected, the Egyptians had a passion for onions and radishes. Herodotus tells us that one hundred thousand workers labored for twenty years to build the gigantic pyramid. If we suppose that every worker ate just one hundred-gram onion per day, the hundred thousand workers combined would have consumed ten thousand kilograms of onions per day. This would come to one hundred thousand kilograms (or one hundred tons) in ten days and three hundred tons within one month. If construction proceeded for six months out of every year, the annual consumption would have equaled eighteen hundred tons of onions.

Since trucks and freight trains were nonexistent at the time, the bags of onions needed to be hauled by boat, then reloaded onto oxen and donkeys. This means that two hundred workers would have been busy, exclusively, with unloading and distributing the fifty-kilogram bags. However, the construction workers most likely did not live on onions alone. We have to assume that each worker devoured about one kilogram (raw weight) of fruit, rice, eggs, and vegetables per day. For one hundred thousand people this amounts to one hundred thousand kilograms per day, or three million kilograms (three thousand tons) per month. Hypothetically, we can add to these three thousand tons the amounts of food consumed elsewhere in Egypt (away from the construction site). This sum can be divided by the acreage of cultivated land in Egypt at the time and multiplied by the number of holidays honoring Osiris and Horus (on which days twice the regular amounts of food were consumed). This kind of arithmetic eventually leads us to the circumference of the earth in pyramid inches, the distance between the sun and Alpha Centauri in cubic meters, and the diameter of the ozone hole, which was gradually expanding because of the gases generated by the onion-munching Egyptians.

Calculations even more absurd than these, with even more hair-raising parallels, have been submitted with regard to the pyramids. One example shall suffice:[1] If we measure the number 666 mentioned in the secret revelations of the apostle John in centimeters from the center of the sarcophagus in the Great Pyramid and adjust the number with the axis of the two air ducts in the King's Chamber, the result will point to the month of July 1987. World War III was predicted to erupt during that month. For some mysterious reason, humanity simply ignored the date.

Those who seek mathematical correlations in pyramids (and other ancient buildings) will find them in abundance. Similarly, the length of the desk at which I am working right now stands in some kind of relation to cosmic measurements. Does this imply that all the studious number crunchers and mathematicians who deduce such bizarre data from the Great Pyramid should not be taken seriously?

The Great Pyramid features measurements that are fairly obvious. They are there, plain and simple, firmly integrated into this monumental building. Language may require some crutches to be understood, even vaguely, by a handful of experts after so many thousands of years; numerical values, however, are timeless—1 plus 1 always equals 2, in any corner of the universe.

Where Did the Meter Originate?

Every architect requires a unit of measurement on which to base his or her plans. One of our common units of measurement, the meter, corresponds to one forty millionth of an earth meridian; this length was determined by an international meter convention in 1875. Ever since, a meter ruler made of platinum and iridium has been kept as a standard at the International Organization for Standardization in Paris.

Precise measurements later yielded some very minute discrepancies in relation to the earth circumference; suddenly, the standard meter was not exactly equivalent to one forty-millionth of an earth meridian anymore. A subsequent international meter convention in 1927 decided to adopt a new standard for the meter, making it equal to the consistently reproducible wavelength of the light of the red cadmium line in dry air at 15 degrees Celsius. But this new unit, too, was displaced by the progress of a modern world circled by satellites. The newest standard defines the meter as equivalent to the wavelength of a specific spectral line of the gas krypton (no. 36, at. weight 83.7, melting point 157.2 degrees Celsius). Whether it is based on krypton, cadmium, or a meter standard made from a platinum-iridium alloy, the meter unit always refers to one forty-millionth of an earth meridian.

An exact knowledge of the earth's circumference was necessary to arrive at the original meter unit. If, three thousand years from now, archaeologists dig up the ruins of the Swiss parliament while looking for a basic unit of measurement, they will inevitably end up with the meter. They will be able to corroborate their finding through excavations of other buildings from the same time period. Maybe some smart scientist will make a sensational discovery: This unit of measurement is equal to one forty-millionth of an earth meridian! Pure coincidence, his colleagues will say, because, after all, this would mean that these strange predecessors, who built their houses with heavy stones, would have known the exact circumference of the earth thousands of years ago!

It is the same story with the sacred yard used in ancient Egypt. It measures 63.5 centimeters, which is equivalent to one-thousandth of the distance that the earth rotates in a second at the equator. (There was also an Egyptian yard measuring 52.36 centimeters.)

Nothing More Than Coincidence?

Coincidence? Most likely, because this would imply that the ancient Egyptians knew the rotation speed at the equator and that they calculated in seconds. The coincidences become amazing, however, when they are compounded into monumental complexes rather than being isolated like lonely monoliths. A mathematically gifted acquaintance of mine summarized the controversial data surrounding the Great Pyramid in an excellent brochure.[2] The following is an excerpt:

- The pyramid is oriented exactly according to the four cardinal points of the compass.
- The pyramid is located at the center of the land mass of the earth.
- The meridian passing through Giza separates the

oceans and the continents of the earth into two equal portions. Furthermore, this meridian is the north-south meridian covering the most land, and also represents the natural starting point for the longitudinal measurement of the entire globe.

- The angles of the pyramid divide the delta region of the Nile into two equal halves.
- The pyramid is a perfect geodesic fix and control point. Napoleon's scientists were amazed to find that all land within visual distance can be surveyed through triangulation.
- The three pyramids of Giza form a Pythagorean triangle whose sides have the proportions 3:4:5.
- The proportion between height and circumference of the pyramid is equal to the proportion between a circle radius and the circumference of the circle. The four side walls are the largest and most conspicuous triangles in the world.
- With the help of the pyramid, one can calculate the volume of a sphere, as well as the area of a circle. It is a monument to the quadrature of the circle.
- The pyramid is a giant sun dial. From mid-October to the beginning of March, the shadows of the pyramid reflect the seasons and the length of the year. The length of the stone slabs surrounding the pyramid is equivalent to the length of the shadow of one day. By observing the shadow on the stone tiles, the Egyptians were able to determine the length of the year down to the 0.2419th part of a day.
- The normal length of the square base amounts to 365.342 Egyptian yards. The number is identical to the number of days in the tropical sun year.

- The distance between the Great Pyramid and the center of the earth is exactly equivalent to the distance between the Great Pyramid and the North Pole and is therefore the same as the distance between the North Pole and the center of the earth.
- The surface area of the four pyramid sides combined is equal to the pyramid height squared.
- The top of the pyramid represents the North Pole. Its circumference corresponds to the length of the equator, and the two are proportionally correct. Each side of the pyramid was measured in such a way that it equaled one-fourth of the northern hemisphere, or a spherical square of 90 degrees. (The circumference at the equator is 40076.592 kilometers; the earth circumference calculated through the poles is 40009.153 kilometers.)

This list of mathematical and geometrical coincidences could be continued effortlessly for quite some time, because smart thinkers have published volumes on the subject, which are always refuted by other, similarly intelligent scholars.[3] Would you like another sample?

The inclination angle of the Great Pyramid was chosen in such a way that the afternoon sun does not spread any shadows between the end of February and the middle of October. This was done for a reason: Re, the sun god, was giving mankind a sign. It is not surprising, then, that the average distance between the earth and the sun is also immortalized in the pyramid. It is exactly equivalent to 108 times the height of the pyramid. Coincidence? Hardly. For the height of the pyramid is related to half the diagonal of the base like 9:10.[4]

A person like me, who has never been particularly fond of higher mathematics, is somewhat confused and helpless in the face of this numerical jumble. I am told that the distance between the pyramid and the center of the earth is the same as the distance between the pyramid and the North Pole. This suggests to me that the planners of the pyramid were aware that the earth is a sphere and that they knew its circumference. If, for example, the pyramid were located in Cologne, Germany, at the site of the cathedral, the distance between the pyramid and the North Pole would not be equal to the distance between the pyramid and the center of the earth. Was the location of the pyramid more than a whim on the part of the pharaoh?

I read that the meridian crossing the pyramid dissects oceans and continents into two equal portions. Isn't it obvious that each half of a sphere is of equal size? Yes, but one half of the sphere may have more water, the other more land. The north-south meridian is supposed to be covering the most land. I took out a large map of the world and spread it on the floor, found myself a ruler, and kneeled down. Apprehensively, my wife asked me if I was planning my next trip. My ruler indeed passed more land than anywhere else in the world when it followed the north-south meridian through Giza. I tried alternatives like New York, Hong Kong, and faraway Lima, Peru. The ruler never covered as much land as it did when it passed through Giza. This peculiar game on the living room floor yielded even more grotesque results when I drew a diagonal. The line across the pyramid from southwest to northeast is the longest possible straight line over land around the globe. I tentatively moved the pyramid to Yemen, to Mexico City, to Central Africa and Honolulu. No other location could match the results I got in Giza.

Construction of the Great Pyramid is said to have started in 2551 B.C., over forty-five hundred years ago. A mere three hundred and fifty years ago, the European explorers discovered South America. Even more recent is the detailed mapping of the world, which only occurred in the last few decades. If we continue the southwest-northeast line originating from the Pyramid, we end up, inevitably, in South America, passing from Recife (Brazil) across the continent to the Chilean coast north of Santiago. Were the unknown pyramid planners aware of this? Did they follow predetermined sites and measurements? Did someone, maybe following an age-old religious tradition, dictate to the pharaoh Khufu that he build his pyramid in Giza and nowhere else? Did the measurements derive from the secret kitchens of the gods?

It would not suffice to say that a mathematical genius in Khufu's times came up with some fantastic angles and triangular shapes which resulted in fabulous calculations on papyri. It would not even suffice if this mathematical superstar prescribed the measurements of each building block down to the millimeter, or if he demanded that the ceiling of the King's Chamber be made from polished granite, one hundred blocks precisely. The pyramid designers had to have specific data beyond pure mathematical knowledge, data about the earth's measurements, circumference, and obliquity of axes. Out of what magical hat did they pull their information? Remember that Pythagoras, Archimedes, and Euclid, the great mathematical thinkers, did not appear on the scene until two thousand years later.

The Big Silence

The archaeological experts detest all the numerical conjecture surrounding the pyramids. Their annoyance at the outsiders and pyramid sleuths is understandable, because the questions raised are either trivial or unanswerable. Unfortunately, it is in the nature of a question to linger, unpleasantly, in the air until it is finally answered. Any large construction project these days employs the services of entire teams of engineers and architects. Yet we are supposed to believe that some Egyptian genius single-handedly invented the pyramid and that the mathematical peculiarities of the pyramid either fell down from the sky or do not exist at all. The objection that the Egyptians practiced on other pyramids prior to the Great Pyramid does not carry much weight, because these "practice pyramids" predate the reign of Khufu by only a few decades. Furthermore, they are a far cry from the gigantic dimensions and the mathematical intricacies of the Great Pyramid.

In an excellent volume about ancient Egypt,[5] Egyptologist Dr. Eva Eggebrecht points out that it has only recently been determined that 8,974,000 cubic meters of construction material were moved in the first eighty years of the fourth dynasty alone. This involved the pyramids of Snefru (2575–2551 B.C.), Khufu (2551–2528 B.C.), Djedefre (2528–2520 B.C.), and Khafre (2520–2494 B.C.). Within these eighty years, 12,066,000 stone blocks were split from the rock, sanded down, measured, polished, transported, and inserted into the appropriate place in the designated building. Daily output: 413 blocks! This does not take into account excavation and leveling chores, the manufacture and repair of tools, the construction of ramps and scaffolding, the general expenditure of materials, or provisions for the multitude of laborers.

Neither the design team nor the architects, contractors, priests, or the pharaoh himself said a single word about the construction of the pyramid. There is not one solitary inscription to tell us how it was done. Dr. Eggebrecht finds the builders' silence regarding the construction of the pyramid completely bewildering in light of the fact that the necropolises were not muted in secrecy. She notes that sacrifices were performed in the burial temples of the kings, and priests came and went. Not one of them left a note that would answer even one of our questions about the construction of the pyramid.[6]

I can counter this silence with a number of possible answers:

- The inscriptions in question have not been discovered yet—or they have already been destroyed.
- Building pyramids was the most ordinary job in the world. No one felt it was necessary to describe it to anyone.
- It was prohibited to document the building of the pyramids. Certain pieces of information were to be purposely kept from succeeding generations.
- Our assumptions are incorrect, and the Great Pyramid was already present as a shining example when later generations built their inferior imitations.

"What you don't know won't hurt you" is a well-known saying, but it does not apply to the Great Pyramid. Everybody acts as if their lives depend on discovering the secrets of the pyramid. Self-appointed pyramidologists, as well as respectable engineers, builders, architects, and archaeologists, have tried to solve the mystery. Well-devised, well-calculated, and intelligent solutions have been offered, and rejected. Professor Georges Goyon, an archaeologist and an expert on the technology of ancient Egypt, did a superb job of refuting all the reconstructive theories proposed and then submitted his own interpretation,[7] which in turn was disputed by Professor Oskar Riedl, who boasted that he was offering the "solution of this age-old mystery without miracles and magic."[8] New theories will continue to be submitted and immediately rejected until an actual text about the pyramid is found which clearly describes the building process. The builders of the pyramid have managed to keep us in the dark for thousands of years.

An uninformed reader may wonder what could be so complicated and unsolvable about the building of a pyramid. Take some blocks and stack them up right? Well, the expert knows that the difficulties are truly monumental. To raise a building this high, certain tools are required—and were required back then—such as ropes, rollers, iron chisels, wood scaffolding, pulleys, draft animals, and sleds. And that is where our problems start.

Building Pyramids without Wood?

Professor Goyon, the archaeologist and expert on the technology of ancient Egypt, insists that any hypothesis suggesting wood as the building material for the scaffolding has to be categorically eliminated. The knowledge we have about ancient Egypt allows us to be unequivocal on this issue: Wood was always scarce in the Nile valley. We have sufficient evidence from artifacts to know that carpenters and joiners used even the smallest piece of wood as effectively and economically as possible.[9]

Tamarisks, willows, acacia and palm trees, sycamores and underbrush grew in Egypt at the time. More resistant woods such as cedar and ebony, which could support heavy loads or serve as rollers for forty-ton monoliths, had to be imported. However, such imports of wood from Lebanon, Syria, or Central Africa were very limited. To transport the wood up the Nile, ships were required—ships made of wood! Did camels and horses pull tree trunks through the desert? Definitely not, since both species were unknown in Egypt in Khufu's time; only oxen and donkeys were available as pack and draft animals.

Were the big boulders, which weighed several tons each, pulled up on the scaffolding with ropes? The experts agree that nothing would have worked without ropes. They must have existed, although no one can prove it. A relief on the sepulcher of Djehutihotep, a regional ruler (around 1870 B.C.), shows 170 men dragging a colossal statue through the desert with ropes. A document from the era of Amenemhet I (1991–1962 B.C.) specifically mentions ropes. Sepulchers from the eighteenth dynasty also show simple pulleys being used to stack stones. This does not prove much, however, because the Great Pyramid was built five hundred and fifty years before the reign of Amenemhet I. If future archaeologists discover faded pictures of a modern construction site with cranes, excavators, and conveyor belts, they cannot automatically deduce that the same equipment was in use five hundred years earlier. Moreover, it is dangerous to apply evidence from the eighteenth dynasty to the third or fourth dynasty, the time of Khufu, one thousand years earlier, because this theory creates an inherent contradiction. One would assume that the building quality would be significantly better with the use of pulleys than without. But just the opposite is true. The technology of the Great Pyramid far surpasses the quality of construction in later pyramids. Nevertheless, we must assume that ropes were used in the building of the Great Pyramid, because nothing could have been moved without them.

How about ramps and scaffolding? Here, matters get much more complicated. One widespread theory, which at first glance seems quite sensible, suggests the following scenario: After excavating and polishing the rock foundation at Giza, the workers assembled the bottom layer, one boulder at a time, to form a terrace, leaving only the entrances to the lower rooms open. They then piled sand all around this bottom layer of the terrace. The giant blocks for the second level were pushed and pulled up the sandy slopes with sleds. When the second terrace was finished, sand was piled up to the top, so that the pyramid could grow, level by level, surrounded by a mountain of sand. Professor Goyon has calculated that with an inclination of ten centimeters per meter and a pyramid height of 146.549 meters, the entire plateau of Giza, within a radius of one and a half kilometers, would have been covered by an enormous pile of sand.

The sandpile theory does not work for practical reasons as well. Hoofed animals and their loads would have sunk into the sand, as would have wooden rollers and sleds. Furthermore, many chores were performed for the temples at the foot of the pyramid. Stonemasons were making large blocks and polishing long monoliths with wooden hammers for the galleries inside the pyramid. They could not have performed these duties inside a mountain of sand.

Of course, there would not have to be a sandpile surrounding the entire building. An enormous ramp would do. This idea was supported by Sir Flinders Petrie, the Englishman who also attempted to reconstruct the Labyrinth, and later, in the 1920s, by the German archaeologist Ludwig Borchardt.[10] What kind of material was this ramp supposed to be made of? Wood is out of the question, not only because it was not available in the necessary quantities, but also because it would never have been able to bear the weight of the stone monoliths, sleds, and laborers. Just imagine an inclining wooden scaffold, a kilometer long and 146 meters high at its peak! Several sleds with colossal stones had to be pulled up the shaky wooden ramp while teams of construction workers descended down a second lane with their empty carts.

So the ramp definitely was not made of wood. Maybe rocks and dried mud bricks? Professor Goyon, our expert on these tricky questions, believes that the angle of inclination for such a ramp could not have exceeded three fingers (0.056 meters) per meter. Such a ramp would only make sense if it was going east, toward the Nile, where the boats were unloaded. Unfortunately, the pyramid site is forty meters above the Nile, which would require the ramp to be longer and higher: almost three and a half kilometers long! The volume of such a hypothetical ramp would have been so great that it would have dwarfed the volume of the pyramid.[11]

Regardless of the kind of material used for the ramp, and regardless of such details as whether the surface was covered with oil or made from moist clay to allow the sleds to glide without friction, the ramp had to be re-thought in its entire length each time the pyramid grew by one step. It had to rise steadily and continuously. It was not possible to switch to a steeper angle along the way. Consequently, the angle of inclination had to be continuously readjusted along the entire ramp, and the ramp had to be resurfaced accordingly, regardless of the kind of surface used. Since the ramp was used all day by thousands of laborers, these adjustments could only be made at night. Under the watchful eyes of Horus!

Hurry! Hurry!

Why the haste? Didn't the pyramid builders have all the time in the world? Why couldn't they permit themselves a few days of rest periodically, so that the ramp height could be adjusted?

Pharaoh Khufu, who commissioned the world wonder, ruled for twenty-three years. It is unlikely that he could have ordered the construction of the pyramid prior to his ascension to the throne. His predecessor, Snefru, was already busy building pyramids. Khufu, like any other mortal, had no way of knowing how many years of life the god Osiris had alloted him. He knew, however, the life spans of his predecessors and relatives. There was not much time to complete this marvelous edifice; it was only natural that the pharaoh would have wanted to inspect the completed pyramid before his death. If we consider that Khufu's reign spanned a mere twenty-three years, Herodotus' claim that the Great Pyramid was built in twenty years appears quite plausible. In reality, however, a twenty-year construction period is less than convincing.

Experts generally agree that the Great Pyramid consists of approximately two and a half million stone blocks. Some of them weigh close to forty tons, or even more, and others only a single ton. The majority are estimated at about three tons. If the pyramid was completed in twenty years, a hundred and twenty-five thousand stones would have been moved annually. We can safely assume that the ancient Egyptians did not work every day of the year. Even without the benefit of unions, there must have been festivals and holidays. Let us assume, then, that the workers labored three hundred days out of every year. One hundred and twenty-five thousand monoliths divided by three hundred working days leaves us with a daily output of 416.6 stone blocks. Such numbers tend to make us generous. I therefore assume that the pyramid workers slaved away twelve hours a day—what a terrible working day!

If we divide 416 stones per day by twelve working hours, we arrive at about 34 stones per hour. Divided by sixty minutes . . . we are talking one huge stone block every two minutes! This simple calculation is based on stone blocks that are all prepared and ready to go—which is misleading. First, the monoliths had to be chiseled out of solid rock, trimmed down to exact proportions, polished, and finally transported to the construction site.

Even with our present-day technology, we would not be able to manage such a feat. This calculation, which reflects rough averages, has been attacked with a number of questionable arguments. It has been argued that work would have been much easier on the lower terraces than on the upper levels. Moreover, fewer monoliths were required as the building progressed. How would that affect our averages? Plus: The higher the pyramid rose, the higher the hypothetical ramp had to reach. The higher it grew, the more work it took to haul the stones to the top of the pyramid. Maybe now you are beginning to understand, dear reader! What organization! What planning! Every two minutes a prefabricated block had to be put in its proper place!

These numbers have not been cooked up by pyramid fanatics. Who would deny the justification of questions being raised about the pyramid?

What Do the Eyewitnesses Say?

Ancient historians wrote not only about the Labyrinth, but about the pyramids as well. Herodotus claims that the pharaoh Khufu forced all Egyptians into labor on the pyramid and that it took ten years just to build the road on which the building material was transported to it. Included in these ten years are "the works on the mound where the pyramid stands, and the underground chambers, which Cheops [Khufu] intended as vaults for his own use: these last were built on a sort of island, surrounded by water introduced from the Nile by a canal. The pyramid itself was twenty years in building."[12]

This account, which Herodotus received from the priests, is followed by a description of the building process:

> The pyramid was built in steps, battlement-wise, as it is called, or, according to others, altar-wise. After laying the stones for the base, they raised the remaining stones to their places by means of machines formed of short wooden planks. The first machine raised them from the ground to the top of the first step. On this there was another machine, which received the stone upon its arrival, and conveyed it to the second step, whence a third machine advanced it still higher. Either they had as many machines as there were steps in the pyramid, or possibly they had but a single machine, which, being easily moved, was transferred from tier to tier as the stone rose—both accounts are given, and therefore I mention both.

These "machines" Herodotus described turned into a hotly debated issue among scientists. Herodotus mentions wooden scaffolds, which were used to move the stones up, level by level; he probably had some kind of lever or pulley in mind. That would seem quite plausible, but unfortunately this possibility is rejected by a majority of experts, people who should know what they are talking about. John Fitchen, a professor of architecture at Colgate University, who has dealt extensively with the construction technology of our ancestors, has the following to say about the Great Pyramid:[13]

> It can be stated categorically that, except for a very few stones of relatively small size (and even these, only in spe-

> cial circumstances), the ancient Egyptians never lifted blocks by means of tackle and pulleys nor suspended them by ropes from above. Their massive, sometimes colossal monoliths precluded the possibility of suspending their dead weight from ropes. Instead, blocks of stone were raised—whether by wedge, lever, or rocker—by jacking operations.

Fitchen's opinion is shared by Diodorus Siculus, who tended to be more pedantic in his writings than his predecessor Herodotus. Diodorus assures us that machines had not been invented in those days. It is interesting to compare texts from the two authors, but one must be aware that both Herodotus and Diodorus merely reported what their Egyptian sources told them. After all, the pyramid, in its entire splendor, had been completed two thousand years before these historians wrote about it.[14]

> The eighth king after Remphis was Chemmis of Memphis. He ruled fifty years and built the largest of the three pyramids, which are accounted among the seven most famous works of the world. . . . It is built entirely of a hard stone which is difficult to work but lasts forever; for although they say no less than a thousand years have since elapsed until our lifetime (or, as some writers have it, more than three thousand and four hundred years), yet the stonework has lasted until now in its original condition, and the entire structure is preserved undecayed. And 'tis said the stone was transported a great distance from Arabia, and that the edifices were raised by means of earthen ramps, since machines for lifting had not yet been invented in those days; and most surprising it is that although such large structures were raised in an area surrounded by sand, no trace remains of either ramps or the dressing of the stones, so that it seems

> not the result of the patient labor of men, but rather as if the whole complex were set down entire upon the surrounding sand by some god. Now some Egyptians try to make a marvel of these things, alleging that the ramps were made of salt and natron and that, when the river was turned against them, it melted them clean away and obliterated their every trace without the use of human labor. But in every truth, it most certainly was not done in this way! Rather, the same multitude of workmen who raised the mounds returned the entire mass again to its original place; for they say that three hundred and sixty thousand men were constantly employed in the prosecution fo the work, yet the entire edifice was hardly finished at the end of twenty years.

According to Herodotus and Diodorus, Pharaoh Khufu ruled for fifty years; according to modern archaeology, he only reigned twenty-three years. A longer reign would be beneficial to the pyramid!

Even Gaius Plinius Secundus, the biggest cynic among ancient historians, who had the added advantage of having read the works of his predecessors, described the Egyptian pyramids—"in passing," as they say. He called the pyramids the results of idle and foolish vanity on the part of the ancient kings and claimed that their only motive for constructing them was a determination not to leave their treasures to their successors, in order to keep the common people from idleness.

Finally, an original reason for the building of pyramids! In spite of his biting sarcasm, Plinius was as much in the dark about the builder of the Great Pyramid as we are now, two thousand years later (*Natural History*, book XXXVI, chapter 17).[15]

> But all of these pyramids, the biggest doth consist of the stone hewed out of the Arabic quarries: it is said, that in the building of it there were 366,000 men kept at work twenty years together: and all three were in making three score and eighteen years and four months. The writers who have made mention of these pyramids, were *Herodotus*, *Euhemerus*, *Duris* the Samian, *Aristagoras*, *Dionysius*, *Artemidorus*, *Alexander Polyhistor*, *Butorides*, *Antisthenes*, *Demetrius*, *Demotelles*, & *Apion*: but (as many as have written thereof) yet a man cannot know certainly and say, This pyramid was built by this king: a most just punishment, that the name and authors of so monstrous vanity, should be buried in perpetual oblivion. . . . but the greatest difficulty moving question and marvel is this, what means were used to carry so high as well such mighty masses of hewn quartered stone, as the filling, rubbish, and mortar that went thereto? for some are of opinion that there were devised mounts of salt and nitre heaped up together higher and higher as the work arose and was brought up; which being finished, were demolished, and so washed away by the inundation of the river Nilus: others think, that there were bridges reared with bricks made of clay, which after the work was brought to an end, were distributed abroad and employed in building of private houses; for they hold, that Nilus could never reach thither, lying as it doth so low under them when it is at the highest, for to wash away the heaps and mounts abovesaid.

Plinius also asserts that a well was buried eighty-six yards deep inside the largest pyramid and that the river was channeled into the well.

The contradictory statements by these ancient historians lead to two basic conclusions:

ously multiplied the total weight placed on the ramp. So it was "Everyone keep on moving, without delay, like clockwork, toward the sun."

The Vienna Rocker

No problem, says Professor Dr. Dieter Arnold, an Egyptologist from Vienna, who presented a rocker, a simple device which could have easily moved the stones up to the next level. Such a rocker is very basic—if it works. When I was a child, I once saw a circus clown rocking playfully back and forth in a rocking chair. His mischievous colleagues sneaked up on him and started to place boards under the chair, alternately in the front and the back. In that tenth of a second during which the rocking chair balanced briefly at the end of its rocking motion, before it tipped back again, they would quickly insert a board under the chair. The clown, who was sitting in the chair reading a newspaper, did not realize that his seat was rising higher and higher—until he finally put down the paper and tumbled from the wobbly wooden tower.

Now imagine Professor Arnold's rocker. He theorizes that a monolith is heaved onto the rocker with crowbars and tied down with ropes. Two workers jump on one side of the rocker, causing the rocker to tilt to one side. In a wink, two different workers slide a board under the rocker. The first two workers jump off, and two others jump on the other side. Quickly, a board is inserted under the other end, and the rocker, with its whole heavy load, has been raised by a few centimeters.

It must have been quite a sight to see! Construction workers bouncing all over, jumping up and down as if they were competing in a jump rope contest! Maybe we should introduce a new Olympic discipline: rocker jumping! Another possibility would be that two workers were standing on top of the rocker, sustaining its rocking motion by shifting their body weight back and forth.

The rocker theory only works with light weights; with heavy loads the rocker would be ineffective. The heavier the stone on the rocker, the thinner the boards would have to be. With a weight of three tons, it would be impossible to insert a board under the curve of the rocker, because it would function like a brake and bring the rocking motion to an abrupt halt. The impact as the rocker hit the side of the board would eventually destroy the bottom surface of the rocker, which after all was not made of steel. Only a very minimal increase in height would be possible with a thin board, which, in turn, would splinter and disintegrate as soon as the total weight of rocker, load, and bouncing workers reached a few tons.

The merry rocking theory is totally out of the question with the use of long, instead of square, monoliths. They could not be mounted lengthwise, in the direction of the rocker, because their ends would hit the ground once the rocker was set into motion. They could not be mounted across either, because they would affect the balance and because there was not enough room. Yet long stone beams were used in abundance in the building of the Great Pyramid. The ceiling of the King's Chamber, with the compartments above it, consists of more than ninety granite beams, each weighing more than forty tons. This theory is truly off the rocker!

Flooding and Lifting

Professor Riedl of Vienna solved the mystery of the pyramids without rockers or ramps, without one hundred thousand workers, and without hocus-pocus! How were the forty- and fifty-ton granite beams transported from Aswan to Giza? On barges? Not a chance, says Professor Riedl. *Under* barges! Riedl remembered the ancient mathematician Archimedes (born 278 B.C.), who invented not only the Archimedean screw but also a number of clever war machines. It is said that this mathematical and practical genius realized one day, while swimming, that his body weighed less in water than on land. We call it buoyancy. At some point, after yet another granite beam fell off a barge into the water, the Egyptian builders must have come to the same conclusion: Stones weigh less under water. Professor Riedl claims that the Egyptians tied their heavy loads under the water surface between two boats. To do that, the ships were anchored and flooded with water until the load was securely fastened underwater. Then, the barges were emptied by hundreds of busy hands so that they bobbed up to the surface, simultaneously raising the granite beams attached to them.

Theoretically, Riedl's suggestion is quite sensible. Whether it would have been feasible on a thousand-kilometer journey up the Nile, through shallow water and rapids, would have to be demonstrated in an experiment using ancient Egyptian barges. The minimum weight for each stone beam would have to be forty-five tons, because the raw monoliths were bigger than the finished beams after they were polished. Once they arrived in

Giza, the barges were directed into a specially designed harbor, where they were flooded once more, causing the load to sink to the bottom. The monoliths, which were still attached to ropes, were then hauled onto the sleds by a crew of workmen. It is even possible that the sleds were placed underwater beforehand so that the stones could be lowered directly onto the sleds.

Professor Riedl does not believe that the sleds were dragged along an endless ramp by hundreds of swearing and sweating workers, but rather that they were moved with the help of stationary winches. He imagines an array of winches positioned all over the plain at Giza. The horizontal turnstiles were driven by men or oxen. The sleds were moved from one winch to the next until they reached the foot of the pyramid, where the stones were loaded onto wooden lifting platforms. Professor Riedl suggests twenty such platforms, each with a length of five meters, on each side of the pyramid.

The principle is simple and works just as easily as the practical devices used by window washers on skyscrapers, without ramps, scaffolds, or piles of sand. Several pulleys are mounted on each completed terrace of the pyramid. The ropes hanging down from the pulleys are attached to a long wooden platform, which in turn, has a crank-operated pulley on each end. If one pulley is operated, the wooden platform is lowered on one side. The stone monolith can be pried off the sled and onto the platform. The load is secured with a stop, a few men turn the crank or turnstile, and the slanting platform is gradually lifted into a horizontal position. Then the pulleys move the load, as well as the workers, up to the next terrace of the pyramid. It reminds me of Laurel and Hardy attempting to paint the outside of a house; their paint bucket slides down the slanting platform and crashes to the ground.

Professor Riedl's theory is excellent. It explains the construction of the pyramids "without miracles and magic"—if indeed the conditions for this kind of building process could be met. Large quantities of wood would have been required for the barges and the submarine transport of the stones, as well as for the countless sleds, winches, pulleys, and lifting platforms. Ultimately, Professor Riedl's theory would have been doomed by the requisite yards and yards of finest-quality rope without which no winch would have turned and no platform would have climbed the pyramid walls. The pyramid builders supposedly had hemp rope. Hemp rope? At best, hemp rope would be able to sustain a load of two or three tons, how many ropes would have been required to lift a fifty-ton monolith? At what point would the rope have snapped off the round wheel? At what point would the thin rope have been shredded on the turnstiles? At what point would the lifting platform have crashed down from the ninety-sixth terrace, splintering the edges of the carefully inserted lower-level stones?

It is unlikely that the construction of the pyramid could have proceeded without accidents; yet there is no indication that crashing boulders damaged any parts of the growing edifice. Did the ancient Egyptians have the necessary expertise about pulleys and these rather sophisticated lifting platforms in Khufu's times (2551 B.C.)? One would think that if the answer is affirmative, the succeeding generations of pharaohs would have had the same kind of technology. Why would Khufu's successors build such pitifully small pyramids if the technology was already available and construction was a cinch with the pulleys and lifting platforms? Consider Pharaoh Niuserre (2420–2396 B.C.), for example, who lived merely 130 years after completion of the Great Pyramid and whose reign was slightly longer than that of his predecessor, Khufu. He had the same amount of time to build his pyramid, and one would think that construction technology had actually improved in the 130 years since Khufu; builders and aritects learn much in such a time span. However, Niuserre's pyramid at Abu Sir measures a mere 51.5 meters in height; the pyramid of his predecessor Sahure (2458–2446 B.C.) is only 47 meters high, which is still four meters higher than the diminutive pyramid of Pharaoh Unis (2355–2325 B.C.), who was also in the fifth dynasty. Egypt offers a great number of pyramids, step pyramids, unfinished pyramids, and collapsed pyramids. Nowhere have archaeologists found even a trace of a rotted platform or a mounting block for a pulley.

Concrete That Lasts for Centuries

Who cares? says Professor Joseph Davidovits, director of the Institute for Applied Archaeological Science at Barry University in Miami, Florida. The Egyptians did not get the stones for the great pyramids in Aswan or in any other quarry. Nor did they hoist them around with pulleys. They poured them at the construction site, just like concrete. Wow!

The evidence submitted by Professor Davidovits, a chemist, has all the trimmings of a mystery novel. It goes like this:

In 1889, the Egyptologist C. E. Wilbour found a stele covered with hieroglyphs on the tiny Nile island of Sehel, north of Aswan. Sehel is one of the few places in Egypt where one can see the ancient Egyptian gods immortalized in splendid inscriptions in the rock. The inscriptions were translated in the last century by the archaeologists Brugsh, Pleyte, and Morgan, and again in 1953 by the French Egyptologist Barquet. The experts agree that the hieroglyphs on the famous "famine stele" date to the era of the Ptolemies (around 300 B.C.), although the text itself recounts a time one thousand years earlier. Of the twenty-six hundred hieroglyphs found on the stele, six hundred and fifty describe the manufacture of artificial stones![19] It is said that Khnum, the ancient Egyptian god of creation, imparted this knowledge to Pharaoh Djoser (2609–2590 B.C.), the builder of the first pyramid, in a dream.

It must have been a strange dream, because Khnum dictated a list of twenty-nine minerals and various natural chemicals to the pharaoh and also showed him the naturally occurring binding agents needed to fuse the artificial stones together. Pharaoh Djoser, the builder of the step pyramid of Saqqâra, was not the only one to receive such divine messages; the same thing happened to his chief architect Imhotep, who later was worshiped like a god in Egypt and whose tomb archaeologists have never been able to locate.

Columns 6 through 18 on the "famine stele" listed the ingredients necessary to make concrete and the sites where these ingredients could be found. Following these divine instructions, Imhotep mixed natron (sodium carbonate) and clay (aluminum silicate) and combined them with additional silicates and aluminous Nile mud. By adding arsenic minerals and sand, Imhotep obtained a quick-drying cement with the molecular structure of natural stone.

At the Second International Conference of Egyptologists, held in 1979 in Grenoble, France Dr. D. Klemm, a mineral expert, astounded the incredulous archaeologists with his analyses of pyramid stones.[20] Dr. Klemm and his scientists had analyzed a total of twenty different rock samples from the Great Pyramid and had arrived at the conclusion that each stone had to have come from a different region in Egypt. If you think that perhaps each Egyptian village contributed a stone to the great building, you are wrong, because the tested samples, individually, contained ingredients from various regions. Normally, a natural granite block has a uniform density; the stones examined by Dr. Klemm, on the other hand, were denser on the bottom than on the top and also contained too many air bubbles.

Professor Davidovits lists two additional pieces of evidence which could, literally, cement his theory.[21]

In 1974, the famous Stanford Research Institute in California joined with scientists from Ain-Shams University in Cairo to perform electromagnetic readings on the Great Pyramid. High-frequency waves were shot through the rocks, on the premise that they would not be reflected completely by dry monoliths. In fact, the scientists were quite certain that the readings would help them discover secret passages and chambers, because the pyramid and the whole Giza plain were considered completely dry.

Contrary to expectations, the readings were chaotic.

The high-frequency waves were totally absorbed by the rock. What was the problem? The pyramid stones contained far more moisture than natural rock. Computers calculated that the Khafre Pyramid alone contained several million liters of water. Professor Davidovits concluded: "The blocks are artificial."[22]

The second piece of evidence could have been taken from an Agatha Christie novel. When Professor Davidovits performed a microscopic analysis of rock samples from the Great Pyramid, he discovered traces of a human hair and later an entire hair, twenty-one centimeters long.[23] How did the hair end up in the rock? Maybe an egyptian concrete mixer lost it.

In the meantime, Professor Davidovits mixed up a variety of Egyptian cements and concretes, based on ancient Egyptian recipes. The "new"—age-old!—concrete is much harder and much more resistant to environmental factors than our modern concrete, because it dries faster and more thoroughly as a result of its chemical reactions. Is it surprising, then, that the company "Géopolimère France" has started to make concrete according to this old recipe? Dynamit Nobel also intends to manufacture the new cement mix, and "Lone Star" has introduced the harder and faster-drying concrete mix in the United States.

Pyramids in the Fog

Once again I was standing on the gentle hill just south of the pyramids at Giza, together with my assistant, Willi Dünnenberger. It was early in the morning, before six o'clock, on May 12, 1988. Our smiling taxi driver Kamal had chauffeured us to the pyramids in the middle of the night because we wanted to photograph the sun rising over the pyramids. It was in vain! Even though the pyramids were barely three hundred meters away from us, we still could not see them one hour after sunrise. The magnificent buildings were shrouded with heavy gray curtains of dense fog, which steadfastly refused to lift. Even at this early hour, we were barraged by chattering tour guides: "Welcome to Egypt!" Only Horus, who sees everything, knows where these pesky pseudo-guards spend the night. They are omnipresent and annoying—around the clock.

We were shivering. Willi inspected the cameras while I trotted off about fifty meters in the direction of the pyramids. Undoubtedly, I figured, the silhouettes of the symmetrical triangles would have to emerge from the fog at any moment. It was eight o'clock now, and the fog was shimmering softly like white cotton candy. Reluctantly, a faint ray of light, pale as the moon, filtered through the steaming mist that tenaciously guarded the view of the pyramids.

"I wonder if they had fog in Khufu's times," Willi asked, and I was wondering the same thing. In that case, the construction workers would not have been able to work in daylight for twelve hours. Finally, around half past eight, the shroud dissipated. Six majestic triangles, two from each pyramid, shimmered in the early-morning light with cool and imposing grandeur. The Egyptians say that man fears time, but time fears the pyramids.

Kamal was negotiating with the bearded supervisor at the entrance to the pyramid. We wanted to get in before the imminent onslaught of tourist buses. We stopped for a long time in the Grand Gallery leading up to the King's

Chamber. It was perfectly silent. The electric bulbs painted the vertical side walls in a soft yellow light. I felt truly dwarfed in this gallery. The gigantic corridor ascending at a slant to the King's Chamber is 46.61 meters long, 2.09 meters wide, and 8.53 meters high. Such measurements are nothing short of astounding! The lower part of the side walls, up to a height of 2.29 meters, consists of polished limestone monoliths; these are followed by seven rows of enormous corbeled beams terraced in such a way that each beam is eight centimeters closer to the center than the previous one. Consequently, the passageway, which is wide at the bottom, converges at the top. The side walls are tapered toward the ceiling, which is made of horizontal slabs and is only 1.04 meters wide. This construction method is reminiscent of that of the Peruvian Incas, whose doors, windows, and corridors were always shaped like a trapezoid.

The Grand Gallery is the most incredible construction achievement in human history. The realization must strike any visitor like lightning that all the theories about the pyramids can be nothing but fragments of the actual truth. To confound all of us know-it-alls even further, the granite beams facing each other along the side walls of the 8.5-meter-high vault are not horizontal but slanted upward at the same angle as the corridor. The beams and slabs are polished so perfectly that it was difficult for us to see any joints even under the scrutiny of our flashlights. This is definitely a place where people are bound to suspect that the builders of the Great Pyramid may have received some outside help from extraterrestrial gods.

We have forgotten how to be humble. We are continually told that mankind is the pinnacle of creation, the most highly evolved species. That is nonsense! Those who have lost the ability to wonder are unrealistic. Reality is superhuman, entwined with spiritual vibrations and interwoven with the next dimensions of the universe.

I estimate that I have studied about sixty books of theories about the pyramids in the past three years. How was the Grand Gallery built? Nobody knows. All we have is gossip and unproven claims, but everybody is trying to throw their weight around as if they know what they are doing. As Oscar Wilde once said, "Blessed are those who have nothing to say and keep their mouths shut."

A Sarcophagus in the Wrong Place

On the south end, the Grand Gallery squeezes into an 8.4-meter-long tunnel leading to the King's Chamber. At first we had to stoop, since the passage was only 1.12 meters high, but after about a meter the low corridor opened up into a small room, more than 3.5 meters high. Three-ton granite blocks originally barred access to the chamber at this point. After three meters, we had to stoop again. Kamal, who had not smiled in a long time, led the way, followed by Willi and me. Maybe it was my religious upbringing, maybe it was a certain respect, or maybe it was the fact that I was standing in the King's Chamber for the first time without other tourists: In any case, I felt as if I were standing in a cathedral. In a north-south direction the rectangular room measures 5.22 meters, from east to west 10.47 meters. The chamber is 5.82 meters high. Given these sizable proportions, I cannot understand why people insist on calling this a "chamber." The walls of this hall consist of five enormous, horizontal—not vertical—granite beams; the floor is covered with granite as well. The walls feel like polished marble. The ceiling, made from pink Aswan granite, consists of nine huge beams, joined together so perfectly that the seams are visible only as a thin, black thread at best. Hidden above the ceiling are five compartments designed to relieve the weight of the structure; they consist of monstrous blocks, each weighing more than forty tons.

Kamal coughed and pointed to the seamless, polished ceiling. "No one has managed anything like this since Khufu," he said.

Willi aimed his flashlight at the ceiling, inspecting every centimeter of the phenomenal structure. "Who came up with the idea that the hollow spaces above the ceiling were intended to relieve weight?" he asked.

Kamal's smile returned. "What else could they be?"

Reluctantly, I interjected, "The compartments above the King's Chamber remind me, spontaneously, of a Shinto temple, of a gate to another world. It seems to me that the archaeologists should quit interpreting them as structural relief. First of all, the compartments are not aligned with the pyramid axis; that is, they are not located under the apex of the pyramid. And secondly—and this appears much more significant to me—this weight-relief theory implies that its builders knew the exact weight of the pyramid. How could that be in Khufu's time? Can you imagine what kind of mathematical knowledge would be required? Even today, we would need a computer to handle such complicated calculations. Would the King's Chamber have collapsed without the compartments above? Hardly. The space above the ceiling could have been covered by granite beams, whose weight would not have been resting on the ceiling. Besides: Where else in the pyramid do we find such compartments to relieve the weight of the structure?

Kamal silently walked to the black granite sarcophagus, which was only a few meters away, adjacent to the west wall of the chamber. Presumably, it used to be located in the center of the room. The sarcophagus (according to Professor Goyon) measures 2.28 by 0.98 by 1.04 meters.

"There are many unresolved questions here," Kamal said. "Supposedly, the sarcophagus was found empty and without a lid—what is an empty sarcophagus good for? Besides, the measurements exceed the dimensions of the corridor that leads up to the Grand Gallery. How did the sarcophagus, which is made from a single stone, get into this room?"

Willi had an answer. "I suppose they built the pyramid around the sarcophagus. The passageways in the Khafre and the Menkaure pyramids are also narrower than the sarcophaguses found there."

Kamal thought for a while. "Maybe so," he finally said. "But it still does not make sense that the Grand Gallery is much, much higher than the ascending corridor. It would have been easy to carry the sarcophagus vertically in the Grand Gallery, but it does not fit either way in the ascending corridor. I mean, it was a waste to build the Grand Gallery eight and a half meters high. Half of that would have been enough to get the sarcophagus through. And if the pyramid was built around the sarcophagus, like you think, why would they have needed the Grand Gallery?"

The facts defy all logic. Some experts have suggested that the Grand Gallery was intended as an elongated, ascending hall through which the priests would march in solemn procession to pay their last respects to the deceased pharaoh. And granted, a solemn procession is very appropriate in such a situation. But first the priests would have had to stoop and crawl, very unceremoniously, up the ascending corridor to reach the Grand Gallery. That does not seem appropriate.

Willi said, "People who build with such mathematical finesse as these architect priests don't do anything unnecessary. Why would they build pseudo-corridors and empty compartments? It would have taken years to build this nonsense, and they didn't have years to waste, considering the tight schedule they were on."

Kamal laughed. "You forget the grave robbers! It was necessary to confuse them."

Willi looked at me, then at Kamal. "Grave robbers," he exclaimed, staring at Kamal across the sarcophagus, which rested between them like a stone bathtub. "Holy Horus! We are talking about the time period of Khufu, twenty-five hundred years B.C. The whole pyramid business started with the step pyramid at Saqqâra. That's a piddling eighty years before Khufu! Where should the grave robbers have come from? The pyramids were impenetrable, like steel vaults."

He is right, I thought, and Kamal must have thought the same thing, because he was stroking his chin pensively. On the other hand, the granite blocks obstructing the entrance to the corridor undeniably must have existed. The ascending corridor and the King's Chamber were sealed off by these massive granite gates. It is enough to drive anybody up the wall! Why would such an elaborate security system have been required? Why did the pyramid need to be sealed if no pharaoh was ever buried in the Great Pyramid? Why did they devise traps and dead-end passages at a time when no grave robber had yet desecrated a pyramid?

Two Contrasting Elements: Vanity vs. Anonymity

The pyramid builders must have known human nature intimately. They must have been aware that scientific curiosity would consume future generations; after all, a thirst for knowledge is an integral part of human intelligence. At some point in the future, the builders must have figured, people would force the pyramid open. Only then were they supposed to find the testament left behind by the ancient builders. But what is this testament? An empty sarcophagus?

Voices, exclamations of surprise, and giggling sounds were buzzing into our solemn chamber. The first wave of tourists was storming the Grand Gallery. We escaped, past sweaty, expectant faces, out into the glaring sunlight. Every last molecule of fog had been devoured by the scorching sun. A papyrus vendor approached us with "Welcome to Egypt!" While we were leafing through the colorful repertory of classical Egyptian motifs and I was glancing rather absentmindedly at the golden symbols before me, a thought crossed my mind. Hieroglyphs! There were no hieroglyphs in any of the halls, chambers, or corridors, or in the Grand Gallery. How could a pharaoh build the most monumental edifice on earth without applauding his own accomplishment? Without

These gigantic sarcophagi are found in the vaults that were chiseled out of the rock. What purpose did they serve?

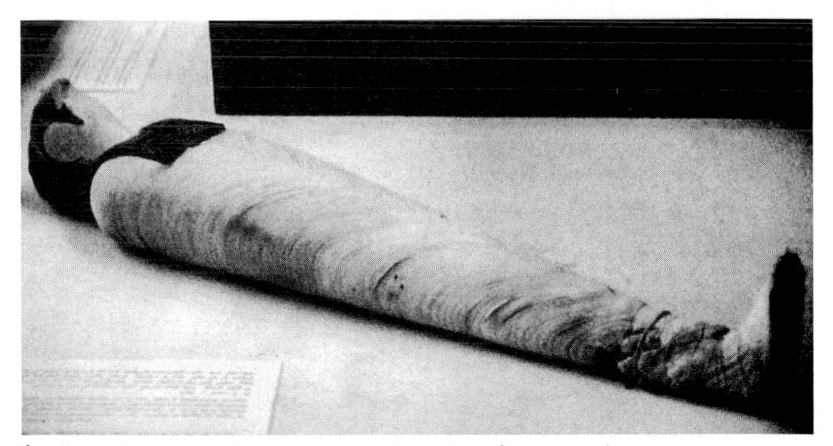

The Egyptian Antiquities Museum is a virtual zoo, with canopic urns and mummified animals, such as falcons and ibises, but also humans.

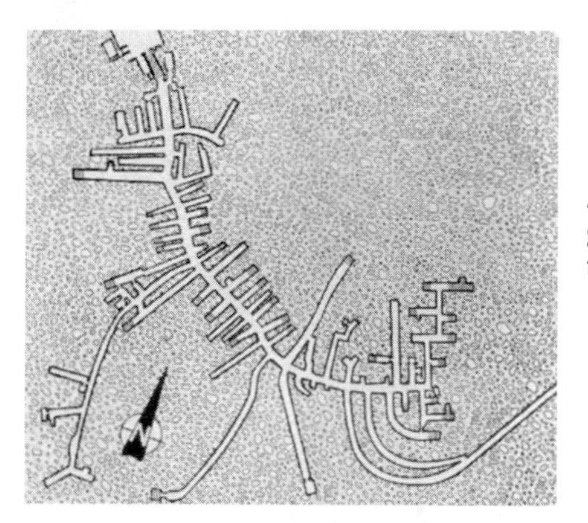

A map of the underground bird galleries of Saqqara, some of them two stories high.

Dr. Walter Brian Emery discovered millions of embalmed birds inside these canopic urns. The pictures were taken in the underground falcon gallery of Saqqara.

Tuna el-Gebel also features a subterranean baboon burial site.

Sumerian seals (*above*) yield myriad mysterious figures. Archaeologists call the winged creatures (*below*, at the British Museum in London) "flying geniuses."

These sphinxes are in the Egyptian Museum in Cairo.

This creature with the snake-like body is also considered a sphinx.

This black obelisk belonging to the Assyrian king Salamaser II is now located at the British Museum in London. It is decorated with unique rectangular reliefs. Details are shown in the next photos.

Even these small human-animal hybrids on the Assyrian obelisk are kept on a short leash by attending guards (*top*). They are animals with human bodies (*bottom*); one of the hybrids is sucking its thumb—the monsters were alive!

All Egyptian gods wear interesting headdresses. The goddess Hathor, for instance, has the horns of a cow and a sun disk. The protrusions above the sun disk are interpreted by the experts as "feathers."

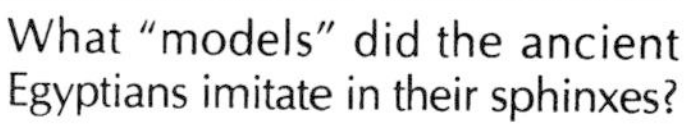

What "models" did the ancient Egyptians imitate in their sphinxes?

Horus, the sky god, is depicted as half-man, half-falcon.

At the top, the goddess Hathor and the ram-headed god Harsaphes; on the bottom, the god Montu, with the head of a falcon, a sun disk, and "feathers."

The pyramid of Pharaoh Amenemhet III at Hawara.

Are these small blocks of rose granite supposed to be the only remains of the huge Labyrinth?

The ramp, made of air-dried bricks, spirals around the pyramid.

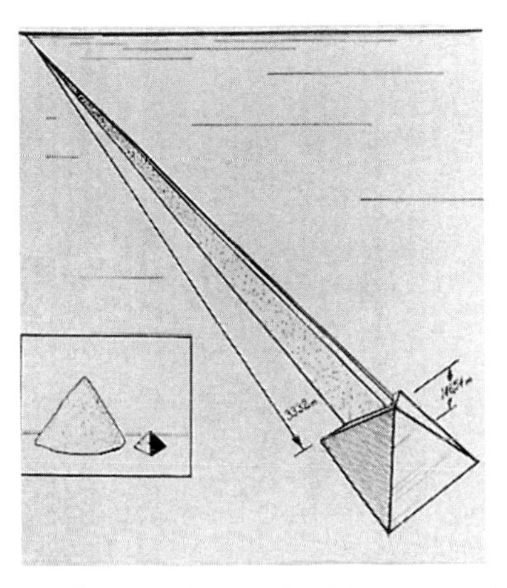

In relation to the height of the pyramid, the ramp would need to be 3332 meters long. Its volume would far exceed the volume of the pyramid.

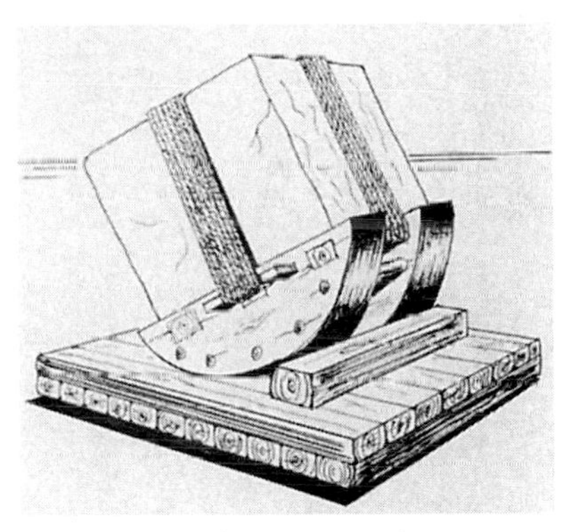

Monoliths on a rocker.

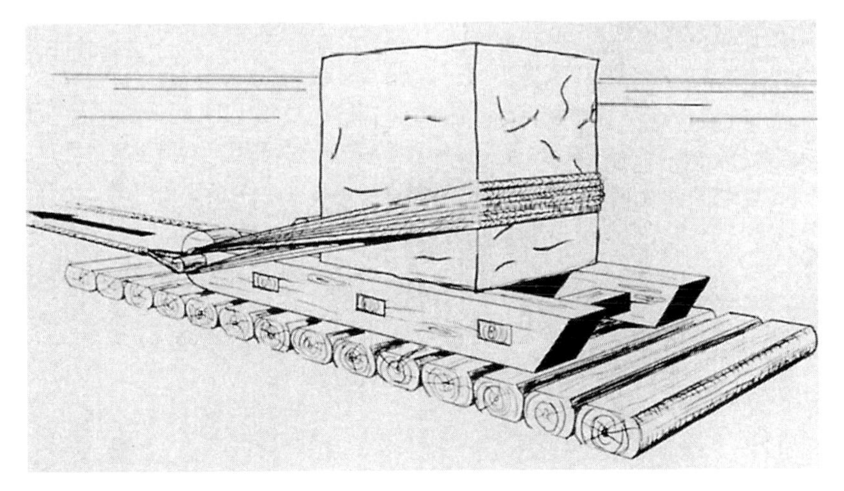

This is how engineers envision the transport of monoliths on wooden sleds.

Inscriptions in the rock on the Nile island of Sehel depict gods from various dynasties.

The winged sun disk in the tomb of Pharaoh Seti.

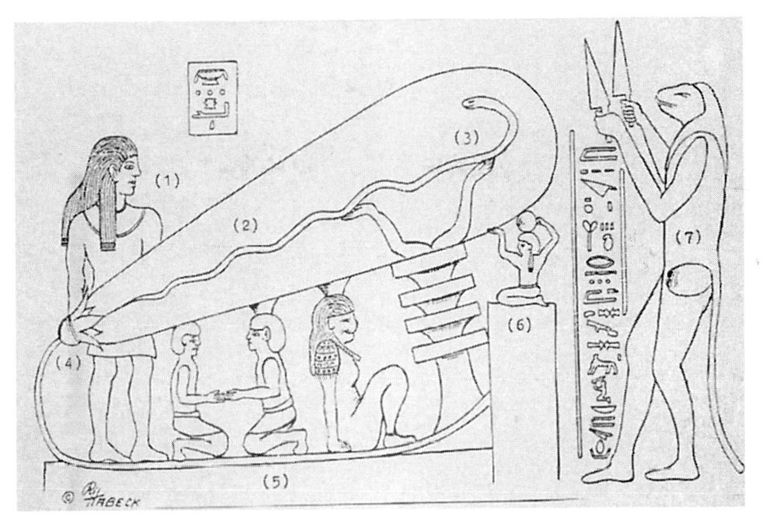

(1) A human holds a bulb-shaped object (2) in which a snake (3) is crawling. The bulb-shaped object and the snake end in a socket (4), from which a cable (5) leads to a small box (6), on top of which the air god is kneeling. The monkey with the sharpened knives (7) symbolizes the danger that awaits those who do not understand the device.

This electric battery is on display at the Museum of Baghdad.

Additional variations on the djed pillar.

The wall reliefs inside the secret crypt under the temple of Dendera demonstrate technical knowledge that was lost over the years.

Djed pillars exist in many variations. An ancient insulating device became a religious symbol for consistency.

The ETORA pyramid on Lancarote.

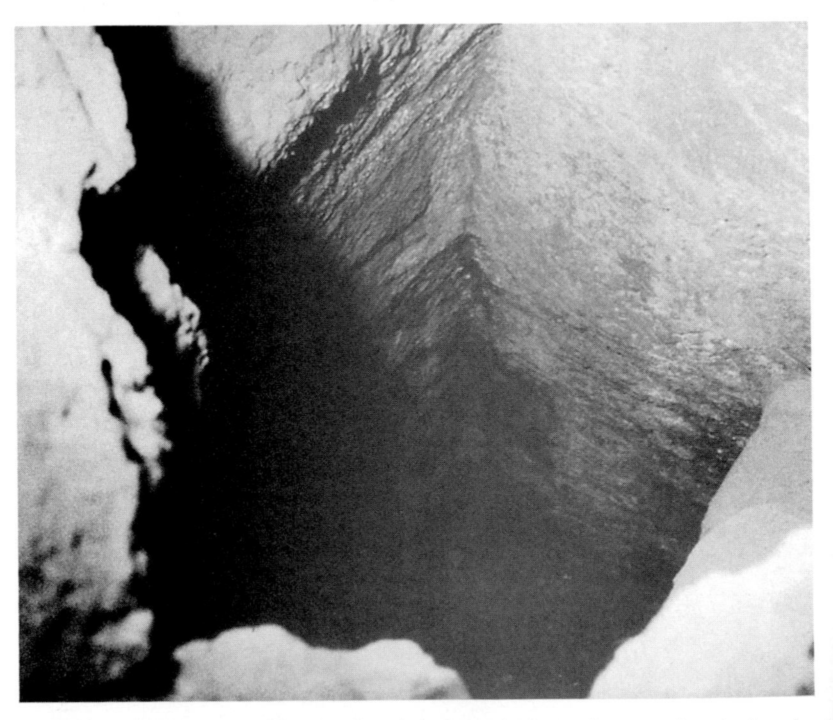

What was the purpose of the dead-end shaft inside the underground rock chamber?

This is the only statue that, without a doubt, depicts Pharaoh Khufu, the builder of the Great Pyramid. It is a puny five centimeters high. (Egyptian Museum, Cairo)

This wall, made of neatly joined monoliths, could just as well be located in the ancient Inca capital of Cuzco in Peru. The picture, however, was taken inside the temple next to the sphinx of Giza. Did the Peruvians have the same architectural teachers as the Egyptians?

The same unnatural vanity to demonstrate their relationship with gods: *Left:* Pharaoh Amenhotep IV with Nefertiti and their three naked children, the smallest on Nefertiti's shoulder, all of them with deformed skulls. *Above:* Deformed skulls at the museum of Ica in Peru.

Rudolf Gantenbrink's robot: UPUAUT-2

immortalizing his name even in the tiniest hieroglyph? This total lack of inscriptions is downright perverse, I mused; the anonymity of the building does not fit the character of the builder.

Plinius felt that it was "a most just punishment, that the name and authors of so monstrous vanity, should be buried in perpetual oblivion." Vanity and namelessness are totally incompatible. If Pharaoh Khufu was indeed a vain man, not to mention a tyrant and oppressor who—according to Herodotus—forced one hundred thousand slaves into labor on the Great Pyramid, accounts of his heroic deeds should have been scribbled all over the walls. It has been argued that the oppressed people knocked off the hieroglyphs praising their dictator. How could they have? And when? The pyramid was completely sealed off; no vandals could have entered it to vent their anger on the pharaoh's inscriptions. Moreover, modern theory holds that no slaves were employed in the building of the pyramid. Karlheinz Schüssler, an Egyptologist, states unequivocally that slavery did not exist in the ancient Egyptian empire.[24]

If this grand edifice was completed without slaves, in voluntary and dedicated cooperation, it is even more unfathomable that there are no written messages anywhere. In a free-enterprise situation, the craftsmen would have made certain to extol their employer's greatness.

Kamal interrupted my train of thought. "Do you know how papyrus is made?" he asked. Between vendors and flocks of tourists, we wiggled our way back to the cab.

"Papyrus isn't made," said Willi, jokingly, leaning from the backseat over Kamal's shoulder. "It grows on the banks of the Nile."

"But how does the plant turn into a pliable sheet of parchment?"

Papyrus: Forever Along the Nile

Willi shrugged. Kamal accelerated and expertly threaded the cab between the clusters of people, camels, and cars on to the highway to Saqqâra. We stopped briefly in front of a carpet-weaving business. Boys and girls, the latter in bright red dresses, were standing in front of a wall, guiding the shuttle through a jungle of warp threads. Boys with pitch-black hair, grayish shirts, and bare feet were operating the creaking wooden looms with swift, sure motions. The children were content. They laughed and sang, and casually thanked us for the baksheesh. Kamal explained that the children design the patterns on the carpets themselves and that they also determine the color combinations. Two kilometers down the road, we visited one of the numerous papyrus factories in the Nile valley. The aqueous papyrus plants, which may grow up to four meters high, have been processed in the same way for thousands of years.

The stalk is cut into pieces of about twenty centimeters. The green bark is then stripped off with a knife. Originally, the elastic bark was used to make belts and sandals, but today it serves as kindling. The white pith inside the stalk is cut into thin strips with a knife and soaked in water for six days. The strips are then flattened in a press or with a rolling pin and arranged crosswise on a cotton sheet. They are next covered with another cotton sheet and pressed again. The sheets are changed frequently until the checkerboard pattern of papyrus strips is all dry. The dry strips stick to each other because the papyrus pith contains gelatin. It takes about six days of drying to produce one of the elastic and rather resilient sheets of papyrus. The finished paper can be painted easily with any kind of paint.

The Egyptians have entrusted their messages to papyrus rolls for thousands of years. How come no word was ever written about the construction of the Great Pyramid? Why is there no reference anywhere to the creator of the most spectacular building of all times? No matter how we look at it, it just does not make sense. Some people have argued that Khufu was not buried in his own pyramid, but somewhere else. Why would he have chosen a different site? His burial vault was the safest in the world. At what point did he decide that he was not going to be entombed in his own pyramid? It is inconceivable that such a decision was made early on in the building process; it undoubtedly would have alienated the priests and architects! Waste all this tremendous effort? Never! With his edifice, Khufu planted a permanent marker in the Egyptian landscape. It is ludicrous to believe that he would have missed such an outstanding opportunity to perpetuate his own glory for all eternity.

These facts leave us with only three alternatives:

a) Khufu's sepulcher was plundered a long time ago.
b) The real burial vault has not been discovered yet.
c) The decision not to be buried in the pyramid was not Khufu's.

I will discuss a) and b) shortly. The third scenario is contradicted by the monolithic reality of the pyramid. The Great Pyramid was sealed completely with huge monoliths and stone gates. The building was entirely finished. If the pyramid had not been completed at the time of Khufu's death and his subjects hated the tyrannical ruler so much that they did not want his mummy inside the pyramid, why would they have finished the building? They would not have lifted a finger after the pharaoh's death. Khufu's successors, after all, had their own construction plans.

Either Khufu is buried in his pyramid . . . or the pyramid does not belong to Khufu.

Pyramid Walls Full of Texts

A mere two hundred years after Khufu's death, Egypt was ruled by Unis (2356–2323 B.C.), the last pharaoh of the fifth dynasty. His pyramid in Saqqâra is only forty-seven meters wide and originally forty-three meters high—minute as pyramids go—but it produced sensational findings.

The walls of the sepulcher, the anteroom, and the entrance to the middle chamber are covered with hieroglyphic texts. The inscriptions are tightly packed in columns, from right to left and from top to bottom. They are the oldest pyramid inscriptions—but not the only ones.

Titi, Pepi I, Mernere, and Pepi II, who succeeded Unis in the sixth dynasty (2323–2150 B.C.), all had the interior walls of their pyramids decorated with inscriptions. In 1965, seven hundred text fragments were found in the Titi pyramid; two years later, French archaeologists entered the Pepi pyramid and discovered walls and corridors covered with hieroglyphs.

In February 1971, the Egyptologist Jean Philippe Lauer and his team excavated the pyramid of Pepi's son Mernere. The light of their lamps fell on massive limestone blocks and the relief of a procession led by a winged genius. In one hand the august divine creature carries a scepter with the animal god Seth on it, in the other the ankh hieroglyph, generally known as the symbol or key of life.

In one of the lower tunnels, the archaeologists climbed over a stone trap that had been triggered by grave robbers and entered two rooms which were separated by huge monoliths weighing at least thirty tons each. The monoliths were arranged in the shape of the letter V, converging on the bottom and reaching toward the ceiling like an enormous victory sign. The stones were decorated with bright white stars, which seemed to be hanging in the room because of the V shape. Some of the walls were inscribed with texts; others depicted mysterious rituals. There were animals, for example, cut in half by a painted line. Archaeologists believe that this was done to symbolically cripple the creatures, to render them harmless.[25] In this way, the deceased pharaoh would not be bothered or attacked by animals on his journey through the realm of the gods. This argument is rather flimsy. If the Egyptians were afraid of animal magic, why would they even have painted animals?

Our thinking has been molded by the old school of Egyptology. In many cases, these previous interpretations may be convincing and correct, but they do not fit our times. Pictures and hieroglyphs which can be interpreted in various ways. Maybe the line cutting the animals in half was not intended to magically cripple them, but to express the fact that the animal was a hybrid: half earthly, half divine.

Those who had hoped that the pyramid texts would shed some light on the construction of the pyramids or on Khufu were sorely disappointed. The hieroglyphs are poetic accounts about mythology, religion, magic, and the preeminent role of the cosmos. The experts agree that the texts, which originated at the end of the fifth and the beginning of the sixth dynasty, reflect fundamental religious beliefs dating back to a more ancient past. It is difficult to believe that the only purpose of the pyramid texts would be to communicate fictitious instructions for a life after death. We consider the laudations and compliments contained in the writings "magical" or "ritual," religious fantasies and aspirations on the part of the pharaoh. The oldest pyramid texts, for instance, describe the pharaoh's desire to meet the sun god Atum-Re in the heavens on his future journey. The scholars assure us that this has to be interpreted in a spiritual sense. Does it? The pharaoh and his priests had very specific ideas about their heavenly travels, although we may consider them childish. They did not travel "by spirit"—they traveled by ship.

Space-Age Technology and Children's Toys

Why do our young ones enjoy playing with model railroads? Because the grown-ups are riding in real railroads. Why does a toddler scoot around the neighborhood in his bright yellow toy car, imitating engine noises and going "beep-beep"? Because his idols drive fancy cars that go "beep-beep." Why do children run around their living rooms with helmets and built-in ear phones, shooting laser guns and playing "conquerors from planet XY"? Because they watch grown-ups on TV doing the same thing. In my book *Habe ich mich geirrt?*[26], I listed a number of cargo cults to demonstrate how native tribes in the present, just like ancient cultures, are imitating technologies they are unable to grasp.

- Inhabitants of the island of Wewak constructed a ghost airport with airplane models made of wood and straw because they were hoping to attract real airplanes in this manner.
- When the inhabitants of the New Guinea highlands encountered white people for the first time in the 1930s, they thought that they were gods. Their mistaken belief was based mostly on the fact that the white people were wearing pants and carrying backpacks. Twenty years later, an eyewitness said, "We thought that they carried their women in their backpacks, and we asked ourselves where the strange creatures left their excrements. Nothing could penetrate their pants."
- The natives of Markham Valley in the eastern highlands of New Guinea made radio stations out of bamboo and insulators out of rolled-up leaves. Tree trunks were used as antennas, and the jungle huts were connected by wires made from twisted plant fibers. Why? Scouts had observed the lifestyle of the white people along the coast.
- When the Russian Maclay arrived with his ship *Vitiaz* in Bongu on the coast of New Guinea in September 1871, the locals watched him skeptically. Once

> they observed him walking around at night with a lantern, and from then on were convinced that he came from the moon. With great difficulty, Maclay explained to them that he was from Russia, not from the moon, but Russia meant nothing to them. In their eyes, the Russian was a very special creature, not only because his skin was white, but because he came in such a large vessel and had appeared so suddenly. Without further ado, the natives declared him a god, Tamo Anut, and his ship became a divine vessel. One day a wooden statue from a shipwreck was washed on shore by the waves, and the natives elected it as the symbol of their new god Tamo Anut.

Similar case histories have filled ethnological volumes.[27] They all reflect the reactions of people faced with a technology beyond their comprehension. It makes no difference if the imitators are children or adults, because grown-ups act just like children when they don't understand a strange technology.

What's New?

Since the beginning of time, and ever since, man has been an imitator. All of us have role models that we secretly strive to imitate. All of us would like to change jobs occasionally or be someone else, if just for a day. We sit behind the wheel feeling like pilots, even though we know that our car is not going to lift off. We slip and slide awkwardly down the slope, wishing that we could be like the elite skiers. We have even fashioned our religious and cultural objects and robes after ancient models.

Our ancestors also copied their knowledge from previous generations. Where is the original crown, scepter, or bishop's staff that our imitations are based on? Who determined that certain rituals can only be performed in specifically prescribed robes? What are we imitating when we carry "heaven" through the streets during a Corpus Christi procession? Why is the sacred Host locked up at the altar? Who gave us angels with wings and radiant halos? Where are the real models for the Ark of the Covenant, the high altar, and the heavenly throne? Where did we get those strange notions of "ascension," "original sin," and "redemption"?

Our present and past knowledge allows us to examine the psychological profile of a pharaoh. He, or his ancestors, had witnessed real gods—extraterrestrials—sailing through the heavens in ships. That was news! Such events must have been the headlines of their oral and written traditions. At first, a handful of humans were chosen to serve these gods. Thoroughly cleansed, of course; dressed in special garments, of course; separated from the divine beings by foyers and barriers, of course. The extraterrestrials did not take any chances of getting infected. Observation and ritual chores and purifications were at the root of the cult, which grew because of people's limited understanding, their natural penchant for imitation, and the gods' miraculous, mind-boggling devices. If it was reported that the gods had sailed through the heavens in ships, then the pharaoh, perforce, had to have a special bark as well. It is irrelevant whether the pharaoh believed he could fly in it after his death or whether he was aware that it was never going to take off. What matters is the motivation behind it.

The pharaohs did not build their sun barks because of a philosophical idea or because they observed the undulating movements of the central constellation. They were imitating something that they were told used to be reality. Humans sailed the Nile with their ships—gods sailed the heavens. People were supposed to believe that their pharaoh was on his way to the gods in his splendid boat, that he was an old buddy and equal partner of the divine creatures. A god-pharaoh and his priests could never have admitted, even if they knew otherwise, that a human's reign ends with death. After all, gods never die.

It is very appropriate, then, that skillfully crafted and ornately decorated sun barks are found next to and under the pyramids. One such boat has been housed in an atrocious-looking building next to the Great Pyramid for years; recently with the help of electromagnetic waves, another one was located in the rocky ground. Such ships appear on temple reliefs from Aswan to the Nile delta. Models are found in museums. Pharaoh Unis—the one with the oldest pyramid texts—also had his own sun bark.

The experts do not know anything specific about the purpose of these boats, even though popular literature wants us to believe otherwise. It is generally assumed that the pharaoh had one ship for the day and another for the night. The Egyptians suspected that the sun moves in the underworld at night, and consequently it was necessary to have two boats. However, the sun barks are also viewed as boats filled with sacrificial offerings, as pilgrim ships, as soul ships, funerary boats, or royal inspection ferries. The pyramid texts which make up the Egyptian *Book of the Dead* permit a number of different interpretations. One example is the "Ode to the Supreme Rule,"[28] in which the poet (or priest?) invokes a goddess named the "Eye of Horus." The goddess is asked to provide water, plants, and food for the pharaoh and to open the gates of heaven so that the pharaoh may travel unhindered. Material nourishment for the fleeting ka and ba?

Pyramid texts 273 and 274 in the pyramid of Unis at Saqqâra praise the heroic deeds which the deceased will perform in space:

> He is the lord of all powers,
> his mother does not know his name.
> The glory of Unis is in heaven
> his power is on the horizon . . .
> Unis is the heavenly bull . . .
> those dwelling in heaven will serve Unis . . .

The inscriptions are even more ambiguous in the actual burial vault of the Unis pyramid. They state that the pharaoh, "like a cloud," is on his way to heaven and that he is resting on a prepared seat in the ship of the sun god. Unis is described as the captain of the sun ship, who pleads for help in the darkness of the universe because "the loneliness is great on the endless trip to the stellar constellations." How true!

A ship conjures up "travels." The pharaohs of the first dynasty undeniably saw themselves as the sons of the gods (just like the Japanese, Persian, and Ethiopian emperors have until very recently). For the son of a god, it was quite natural to return after death to his father, who was busy managing things up in heaven while his offspring was ruling on earth. On our planet, the firstborn prince inherits his father's kingdom, and the same was supposed to happen up above. The pyramid texts extol the deceased pharaoh as the new ruler among the stars, as a potent executioner and judge, to be feared by all the spirits and old gods.

All of this is correct and hardly disputed by the experts, but the Egyptologists fail to perceive any real, practical value in the heavenly boats. Just remember the cargo cults. Symbolic objects are imitations of technology. No pharaoh had the courage to appear before his heavenly father equipped with nothing but ka and ba. He had to bring along his treasures for sacrifices, and as bribes in particularly difficult cases. Real values in a real vessel. The child of today's oil sheikh zooms around the palace in his battery-powered toy Rolls-Royce. The son of the heavenly gods zoomed around in his gilded sun bark.

Astronauts in Ancient Egypt?

A similar interpretation is suggested by another decorative symbol, which appears on every ancient Egyptian temple and monument: the winged sun disk. In the fifth dynasty and thereafter, a golden disk or a sphere with colorful, spread wings symbolizes the sky-ruling falcon and the sun. However, the idea for this motif, which covers entire temple ceilings and countless temple entrances, dates back to prehistoric times, because a decoration from the first dynasty already shows a sun ship on a pair of wings. After the notion of a sun ship gliding along on wings ceased to have meaning, the pair of wings became joined by a golden disk. The image, which appears with geometrical precision above the entrances of halls and vaults, is often accompanied by inscriptions identifying it as "hut" or "api." The root word of "hut" means roughly "to spread" or "to stretch," while the word "api" simply means "to fly."

The winged sun disk is associated with the god Horus, whose center of worship was located in the gigantic temple of Idfu, on the western bank of the Nile between Aswan and Luxor. The present-day temple area, which is still very extensive, is very different from the ancient Horus temple. Inscriptions and archaeological excavations indicate that this complex was built on the ruins of an ancient holy site dedicated to Horus. Similarly, the legend of the winged sun disk, which was inscribed in one of the temple walls in Idfu, derives from ancient sources. The story describes how the god Re and his entourage landed west of this area, east of the Pechennu Canal. Apparently, Re's earthly representative, the pharaoh, was in trouble and had asked the divine sky-traveler for assistance in the fight against his enemies.[29]

> The Holy Majesty of Re-Harmachis spoke to your holy person Hor-Hut: O, you sun child, you exalted one, whom I have created, speedily strike down the enemy who is before you. Then Hor-Hut flew up to the sun in the shape of a large sun disk with wings. . . . When he saw his enemies from aloft in the sky . . . he attacked them so fiercely from the front that they could not see with their eyes, nor hear with their ears. In a very short time, no living body was left. Hor-Hut, shining in bright colors, returned to the ship of Re-Harmachis in his shape as a large, winged sun disk.

Illogical Logic

One must interpret this symbolically, I am told. (It never ceases to amaze me what kinds of things one "must" do.)

Undeniably, hieroglyphs leave much leeway in their interpretation. Many years before Jean-François Champollion, the well-known hieroglyph translator, William Warburton (1698–1779), bishop of Gloucester in England, studied the Egyptian symbols and the ancient tales very thoroughly and recognized that the ancient Egyptians were using two different types of symbols: one kind to reveal the information they wanted to reveal and a different kind to conceal certain things.[30]

That is a fact. But today hieroglyphic texts are served up as an unambiguous certainty, even though interpretations could be as multicolored as the rainbow. More recently, ancient hieroglyphs have been discovered which not even Champollion's translation theory has been able to crack. I find it difficult to accept the legend of the winged sun disk as an abstract tale, somewhere in the foggy realm of ancient religion. After assisting the pharaoh in the battle against his enemies, the flying god Re announces matter-of-factly: "This is a nice place to live." He then names the surrounding regions and praises the gods of heavens and the gods of the earth. We should be allowed to read the original texts more, instead of being told what they mean.

> Hor-Hut, he flew up toward the sun as a large, winged disk. For this reason, he has been called lord of the sky from that day on. . . .

The inscriptions at Idfu prove positively that the winged sun disk was worshiped because it represented divine intervention, not, as we are led to believe, because of the sun's existence in an imaginary world above and below. The Idfu text is very clear:

Harmachis sailed in his ship and he landed near the city Horus-Throne. Thoth said: "The sender of rays, who is conceived by Re, has defeated the enemies in his form. From this day on may he be called the sender of rays, who has been conceived by the mountain of light." Harmachis spoke to Thoth: "Mount this sun disk on all the sites of the gods in Lower Egypt, on all the sites of the gods in Upper Egypt, and on all the sites of the gods."

I would like to mention, by the way, that the term "sender of rays" is not my own invention, but was used, in its German equivalent, by Professor Heinrich Brugsch, who translated the Idfu text in 1870! What has happened to the winged sun disk in modern Egyptology? It has been turned into ceremonial mumbo jumbo. The original image, a sun bark with wings, has been lost completely. The imagination of scholars, unable to grasp ancient reality, turns realities into myths. And the world is put straight again. But whose world?

One particularly sensitive Egyptologist felt that the thought that some god actively intervened in human battles was unbearable, just as unbearable as my notion that extraterrestrials interfered in earthly affairs. Human logic is very fickle indeed. In the Old Testament, for example, God, who descends with smoke, fire, earthquakes, and noise, often joins his chosen people in battle. For real. It is considered logical there. Logical to whom?

Fiat Lux!

While the pyramid texts have been able to shed some light on the ancient Egyptians' simple line of reasoning, they have not been able to "enlighten" us. How did the

Egyptians light the inside chambers of their pyramids? The countless hieroglyphs and artistic expressions on the walls definitely could not have been created in the dark. Were the decorations engraved on the monoliths while they were still outside, before they reached their final destination in a lightless vault? Possibly. To transport the stones, the construction crew would have to have wrapped the decorated walls and slabs in cotton; they could not afford to bump against anything. It is also possible that work was under way while the pyramid was still open, uncovered, and that the vaults were not sealed until the sculptors completed their delicate inscriptions. The lighting problem can be solved for aboveground pyramids, but not so for underground tunnels. Many pyramids are built on deep, excavated foundations, and the tombs in the Valley of the Kings near Luxor consisted of convoluted shafts which no sunlight ever touched. So how did the Egyptians light the walls and ceilings in these colorfully decorated sepulchers? Did each sculptor have his own torchbearer? Did the artists work by the light of oil lamps and wax kettles? Or did the builders set up a system of mirrors to coax sunlight into the dark dungeons?

These are some of the questions Peter Krassa and Reinhard Habeck posed in their well-researched books.[31] It is a lively, intelligent, and irreverent work that should not be missing from the library of anyone interested in Egypt. Krassa and Habeck point out that torches, oil lamps, or wax expel a black smoke and that consequently, we should find soot residue on the walls and ceilings. But there is none. Is it mirrors then? The iron mirrors available at the time were very rudimentary. They lost at least a third of the light with each reflection because of dispersion and absorption. After three mirrors, there would have been darkness.

"It is better to light a small light than to complain about the great darkness," Confucius once said.

Imagine Cleopatra guiding her Roman friend Julius Caesar through the dark tunnels of the pyramid. Suddenly, a mysterious light shines from her hand, lighting up the walls and blinding the eyes of the startled Roman emperor. "What is this magical light you possess, my dear?" asks Caesar apprehensively.

"We call them flashlights," she replies, flattered. "Our ancestors have been using them for thousands of years. Don't you progressive Romans know this source of light?"

Krassa and Habeck summarized their electrifying ideas in an article[32] in *Ancient Skies*, the official publication of the Ancient Astronaut Society: The ancient Egyptians had electricity!

Crazy? This claim can be substantiated rather easily. Our history books tell us that the effects of electricity were unknown until 1820, when they were "discovered" by H. C. Oersted, a Dane. Michael Faraday continued those studies, and Thomas Edison developed the incandescent lightbulb in 1871.

Thomas Edison Was Not the First

This historical interpretation is wrong. The National Museum in Baghdad, Iraq, displays a device consisting of a terra-cotta vase, eighteen centimeters high; a slightly shorter copper cylinder; and an oxidized iron rod with traces of bitumen and lead. This strange vase was unearthed in 1936 by a German archaeologist, Wilhelm König, during his excavation of a Parthian settlement near Baghdad.

König suspected that the curious apparatus could be some kind of energy-producing battery. Ensuing studies confirmed his suspicions. Inside the vase, a thin sheet of copper had been molded into a cylinder with a length of approximately twelve centimeters and a diameter of about two and a half centimeters. It had been welded with an alloy of pewter and lead. The cylinder bottom forms a tightly sealed copper cap insulated on the inside with bitumen. At the top of the vase, the cylinder was also plugged with bitumen. An eleven-centimeter-long iron rod protruded from deep inside the cylinder through the bitumen at the top of the cylinder, insulated against the copper. If the cylinder was filled with an acid or lye, it created a galvanic element, present in the identical combination which Galvani used for the battery named after him.

In 1957, F. M. Gray, an American scientist at the high-tension lab of General Electric in Pittsfield, Massachusetts, proved that an electric current indeed flowed through the device and that the electricity was actually used. He succeeded in producing electricity with an exact replica of the apparatus and a copper sulfate solution. Thus he proved that the artifact found among the ruins of Chujut Rabuah, as well as similar devices discovered in Seleucia along the Tigris and in neighboring Ctesiphon, was an electrical battery. Were similar devices used by the Egyptians?

Ancient wall reliefs in a subterranean crypt in Dendera, seventy kilometers north of Luxor, confirm the claims made by Krassa and Habeck. The temple complex of Dendera is dedicated primarily to the goddess Hathor. In ancient times, she was considered the sky goddess and mother of the sun god Horus. Because the Egyptians saw a gigantic cow image in the night sky, the goddess Hathor was portrayed as a cow in addition to her human form. As a human, she was always depicted with the horns of a cow and a sun disk. She is the goddess of dance, music, love, science, and astronomy.

Light for the Pharaoh

We know from mastabas that Dendera, the temple of the goddess Hathor, was known in the Old Kingdom. Over the course of time, the temple complex lost some of its significance, until it was restored and rebuilt during the Ptolemaic period. Today, the temple structures are definitely worth visiting. Colonnades, walls, and ceilings grant us insights into the religious ideas of the Egyptians of the Greco-Roman period, which, of course, were based on more ancient models. Dendera is also the only site in Egypt with a complete zodiac of the thirty-six *decans* of the Egyptian year. The exquisite relief, with its twelve main figures and mathematical and astronomical signs, was removed from a temple ceiling in Dendera in the last century and sold to King Louis XVIII for one hundred and fifty thousand francs. It can be viewed at the Louvre in Paris. Astronomers who studied the zodiac of Dendera dated it to 700 B.C.; others placed it around 3733 B.C.

Dendera is also unique because of its subterranean chambers and their mysterious wall reliefs, reminding us of long-forgotten times. One of these vaults measures about 4.6 by 1.12 meters and can be accessed only through a small opening. The room has a low ceiling. The air is stale and laced with the smell of dried urine from the guards who occasionally use it as a urinal.

Krassa and Habeck describe the walls as "decorated with human figures next to bulb-like objects reminiscent of oversized light blubs. Inside these 'bulbs' there are snakes in wavy lines. The snakes' pointed tails issue from a lotus flower, which, without much imagination, can be interpreted as the socket of the bulb. Something similar to a wire leads to a small box on which the air god is kneeling. Adjacent to it stands a two-armed djed pillar as a symbol of power, which is connected to the snake. Also remarkable is the baboon-like demon holding two knives in his hands, which are interpreted as a protective and defensive power."[33]

The experts, who should know, are at a loss to explain the reliefs in this small, lightless chamber. They call it a "cult room," a "library," an "archive," or a "storage room for storing cult objects." A "storage room" or a "library" that is accessible only through a tiny opening? Ridiculous! The pictures on the walls are similarly mystifying to the experts. What is a "djed pillar"?

- A symbol for continuity
- A symbol for eternity
- A prehistoric fetish
- A leafless tree
- A notched pole
- A fertility symbol
- The symbolic ear of a plant

Krassa and Habeck, more in tune with reason, interpret it as an insulator. Why not? The Old Kingdom had priests of the "venerable Djed;" even the main god Ptah was called "venerable Djed."[34] A specific rite was performed in Memphis to erect the djed pillar. It was executed by the king personally, with the assistance of priests.

A djed pillar was something very special. Only certain insiders were allowed to handle it. Djed pillars were found even under the oldest pyramid, the pyramid of Djoser in Saqqâra. Looking at the melodramatic interpretations of this curious object, someone like me is bound to start chuckling. What else are we going to construe in our heads before we truly open our eyes and see things the way they are? Hidden way back in the abstruse thoughts of well-respected scholars, the beliefs of ancient Egyptians are reconstructed, while out in the open, in the reality of our century, cargo cults crop up all over. It is blatantly obvious that the djed pillar is a representation of misunderstood technology; even the deaf can see it and the blind can feel it. How did the prophet Isaiah put it in the Old Testament? "Shut their eyes; lest they see with their eyes."

The walls of the crypt below Dendera celebrate a secret science: the knowledge of electricity. I do not expect the experts to share my opinion that the ancient Egyptians used electricity. But it is a shame that, according to Goethe, intelligent people least abandon their sound judgment when they are wrong.

Pyramid Magic

I am standing in an eight-meter-high pyramid, oriented precisely according to the four cardinal points. Above me, four light gray triangles converge at the top of the pyramid. Beige wall-to-wall carpeting covers the floor. Purple pillows are scattered across the floor like patches of flowers. A handful of people, men and women, are sitting on some of the pillows, silently meditating. I scan the pyramid walls with my eyes. Eight small windows have been inserted at the bottom of each triangular wall, for a total of thirty-two windows. My feet are resting on a six-point golden star built into the floor.

A small glass pyramid is sitting in each corner of the pyramid. The inside of the pyramid, faintly lit, is bathed in gentle yellow tones. Someone closes the foam-padded doors. The music begins. At first, it is a calm rustling sound, a distant melody playfully enveloping me like a murmuring brook; then it becomes a roaring and trembling that vibrates from every pyramid surface, flooding my senses and carrying me off into a broiling universe of vibrations. I am enchanted, unable to move away from my star, as Anton Dvorak's *New World Symphony*, performed by the Vienna Philharmonic Orchestra, penetrates every fiber of my body. I remain motionless, frozen, lost in thought, even after the music culminates in a furious crescendo. The sudden silence is like a shock. I feel as if my brain is being dragged through a car wash. A thousand different ideas and inspirations flash through my head, stir me up, pull me away from this world, out into the star-spangled night sky.

Never have I been more aware that the phrase "God is dead" could only have arisen from egocentric brains. This God, who they believe is dead, is all around me, in every molecule, every atom of my existence. While my body is still attached to the floor of the pyramid, my mind is soaring above the top of the pyramid. I perceive myself as a part of the universe, as a flash of lightning spreading in all directions at the speed of light. I have no eyes, yet I recognize the milky-white light falling on the pyramid below. I have no ears, yet I hear, with sweeping intensity, the flowing melodies of *Glass Works* by Philip Glass which now surround the pyramid. In the same instant, I realize with astonishment that I cannot possibly know the title of the composition, that I have never in my life heard of a composer named Philip Glass. What is happening? Where does this supernatural awareness come from which permeates everything and is in all places at the same time? Did somebody spike my drink with a drug? Am I the victim of a spiritual force reaching out for me?

I dive back down into my body, trembling like a wet poodle, and quietly leave the pyramid. Outside, I meet the sound technician, the young man who installed the quadriphonic system inside the ETORA pyramid on the island of Lancarote. ETORA is an esoteric convention center. I had been invited there for some lectures. It is a paradise without any mosquitoes or other pests.

"What is the name of the piece that is playing in the pyramid right now?"

"*Glass Works*, by Philip Glass."

"I must compliment you on the acoustics. You must have measured everything very carefully."

The sound technician laughed. "Nothing was measured at all. I rely on my ear. Besides, there is the pyramid effect."

The Pyramid Effect

The discovery of this effect sounds like a sentimental fairy tale.

Once upon a time on the flowery Côte d'Azur in Nice, there was a hardware store owner named Antoine Bovis. But Monsieur Bovis aspired to something higher than selling screws and nuts. He was a relentless tinkerer and inventor, and even in the 1930s, when no one had heard the term "New Age" yet, Antoine Bovis was the leader of an esoteric circle.

Not surprisingly, Monsieur Bovis's store sold more than iron bars and tools; it sold special magnetic pendulums, "biometers" invented by Bovis himself, and various devices. During a trip to Egypt, which, among other places, brought him to the Great Pyramid of Giza, Bovis made a curious discovery, one that other tourists passed by without a second thought. A tiny desert mouse was lying dead on the floor of the King's Chamber. Bovis speculated on how this small creature had entered the age-old building.

Antoine Bovis gently nudged the dead mouse with his foot. He wanted to see if bugs or ants had found their way to the animal corpse. Monsieur Bovis searched the floor carefully for clues, turned the mouse over, again and again, then finally bent down to pick it up. The realization struck him like lightning: The desert mouse was light as a feather, shrunk and mummified.

What mysterious forces were behind this discovery? Why did the dead mouse show no signs of decay?

Upon his return home, Monsieur Bovis constructed a pyramid model out of iron bars and wood. Light as it was, the dead mouse in the Great Pyramid weighed heavily on his mind. Intuitively, he did the right thing from the very beginning: Antoine Bovis arranged his pyramid in a north-south direction, just like the original pyramid in Giza. He then placed a small wooden podium inside the pyramid, one-third the height of the pyramid. The podium was intended to represent the site of the King's Chamber, which is one-third above the foundation of the Great Pyramid. Finally, following his intuition, and the fact that beef stew was planned for dinner, Antoine Bovis placed a small piece of beef on the podium.

Ordinarily, the meat should have started to stink after a few days, but it did not. It became increasingly drier, more and more desiccated, as if an invisible force was drawing all fluid out of the beef chunk. Bovis observed the mummification with irritation. Then he initiated a whole series of experiments with and without the model pyramid.

All organic substances inside the pyramid were dehydrated; those outside the pyramid rotted.

"Sounds logical to me," I said, when I read the story for the first time. The meat inside the pyramid is almost hermetically sealed from the outside. As with vacuum wraps, bacteria cannot enter. Buy why does the meat dry out? What is drawing out the juices?

Czech Patent No. 93304

Similar thoughts must have crossed the mind of the Czech radio engineer Karl Drbal, who read about Monsieur Bovis's experiments in an obscure magazine. Drbal reproduced Bovis's experiments and found that the results were the same. But he felt that meat, eggs, and cheese were the wrong ingredients in these pyramid experiments. How about inorganic, "nonliving" substances? Could a rock, a spoon, or a thimbleful of water be dehydrated in a model pyramid?

Karl Drbal looked for a small object which would fit in his tiny eight-centimeter-high cardboard pyramid (baseline: twelve and a half centimeters). He happened upon an old razor blade whose life had extended beyond all usefulness. The radio engineer suspected that the blade would lose its last spark of sharpness inside the pyramid. Twenty-four hours later, he examined it under a magnifying glass. Was he hallucinating or did the blade indeed look as if it had been sharpened? Without a moment's hesitation, Karl Drbal shaved off his beard stubbles with the old blade. Then he returned the razor to the pyramid. He figured that there had to be some way to finish off the wafer-thin metal. But the next day he got a perfect shave again from the old blade. What was happening? Was it a figment of his imagination or was the blade getting sharper every day? He slowly stroked his smoothly shaved chin, which showed no trace of even the tiniest cut. Amazed, Drbal returned the blade to the pyramid—and continued to shave with it without incident for fifty full days.

All of this happened in February and March of 1949. The radio engineer tenaciously continued experimenting for another five years and three months, until July 6, 1954, averaging 105 daily shaves with each blade. Overall, Drbal used eighteen different blades from various manufacturers; the final number of shaves accomplished with one blade ranged from 100 to 200, with daily use.[35]

Karl Drbal continued to use his free razor sharpener even after the termination of his experiments. Over a period of twenty-five years he only needed an amazingly few twenty-eight razor blades! Understandably, razor manufacturers were less than thrilled about his findings.

The next stop, obviously, was to apply for a patent. But how? Karl Drbal had no idea what process was responsible for the goings-on inside the model pyramid. He applied for a patent anyway, and because he knew that the patent committee was hardly going to be convinced, he presented the metallurgist on the committee with a small, complimentary pyramid containing a razor blade. The metallurgist, though skeptical, tried the pyramid blade on his own beard.

In the summer of 1959, Karl Drbal was granted a patent for a "device to maintain the sharpness of razor blades and razor knives." Czech patent No. 93304.

Since then, the razor blade experiment has been repeated thousands of times, with identical results, as long as the model pyramid and the cutting edge of the razor blade have been oriented precisely in a north-south direction. In his TV program *Terra X*, Dr. Gottfried Kirchner reported on a strictly scientific experiment conducted by Professor J. Eichmeier of the Technical University of Munich. As part of the experiment, half of a razor blade was placed in a Plexiglas pyramid for eight days, the other half in a closed drawer. Both halves were then examined under a surface electron microscope. According to Dr. Kirchner, the differences in the width of the cutting edge, as well as in the surface structures of the two blade halves, were significant.[36]

To Explain the Incomprehensible

What force affects the molecular structure and thus the composition of atoms in a steel blade? Why does the experiment only work with a pyramid, but not with a cube or a cylinder? What is so special about the pyramid shape, and why does the mysterious energy flow only if the pyramid is oriented in a north-south direction? By now, it is indisputable that changes occur not only in steel, but in other materials as well. Yet no one has a satisfactory explanation for this phenomenon. Dr. Kirchner refers to American scientists who believe that the radiation energy of the objects is being contained inside the pyramid. The energy cannot escape through the side walls, but instead is reflected inside the pyramid. The constant reflections, according to this theory, change the structure.

At first glance, this may sound sensible, and yet it raises more questions than it answers. Every molecular compound, i.e., all matter, radiates. Radio astronomers were able to prove the existence of many organic and inorganic substances in space solely based on the radiation emanating from them. Radiation also means loss of energy. If a radiation source expended all of its energy, it would cease to exist. In the subatomic range, the lost energy is constantly replenished because electrons, as the components of the atom, alter their state and in essence jump from one energy level to the next. It is just as easy for an electron to slip through the cardboard wall of a pyramid as it is for air to pass through the mesh of a fishing net. How could the angle of inclination of a pyramid possibly interfere with that?

Karl Drbal identifies a number of different reasons for the pyramid effect. So-called dipolar water molecules are stored in the tiny crevices of the crystalline structure of the razor blade edge. The resonance of the radiation energy forces them out of their hiding spots. Symbolically, according to Drbal, the process could be called the dehydration of the razor blade edge.

Where do these dipolar water molecules magically disappear to? After all, they are supposed to be reflected by the inside of the pyramid. Drbal maintains that they attach themselves to the air surrounding them, which, in all likelihood, is the only possible solution. The experimental pyramids are air-permeable. But what happens if the pyramid forms a vacuum that does not allow air to enter or exit the pyramid? What measurable forces are necessary to eject or release the dipolar water molecules from the steel?

The Soviet physicist Malinov cited "electromagnetic waves" in connection with the magnetic fields of the earth for the curious pyramid effect. Buy why, for the sake of all pyramid-building pharaohs, do these waves kill mold- and decay-producing fungi and bacteria in food while at the same time preserving the foods and even enhancing their natural flavor? The Ancient Astronaut Society, a nonprofit organization dealing with my theories, wanted to find out more and encouraged its members to conduct pyramid experiments with all kinds of different materials.[37] Over the course of several weeks and months, it received 118 reports, from men and women with various professional backgrounds and from students. All of them had crafted model pyramids of various sizes from a variety of materials, had stored them in gardens, basements, attics, bedrooms, on an air mattress anchored in a swimming pool, and even in a refrigerator, and had equipped them with the most amazing objects. One sixteen-year-old boy from Holzkirchen in Bavaria reported that he had placed ants in a small plastic box and that they died after four days. Another sixteen-year-old described his experiment with flies, which died after twenty-four hours. Obviously, the poor creatures lacked oxygen, liquids, and food. I immediately called the young scientists to tell them to terminate their horrifying experiments at once.

A female teacher who was vacationing in Ticino, the southernmost canton of Switzerland, placed a piece of slightly moldy bread inside her pyramid, which was covered with wax paper. She then deposited her pyramid, twenty-two centimeters high, in the basement because "it is so humid there, and mold fungi love humidity and darkness." Eighteen days later, the mold was gone. The bread crumbled into tiny morsels. Pow!

A retiree from Arbon, on the shores of Lake Constance, got a surprise when he placed a flat, round candle, the type that is used in potpourri burners, inside a glass pyramid. He reported that he merely wanted to see if the flame would burn steadily. The flame, however, kept going out because of a lack of oxygen, and the sixty-eight-year-old man lost interest. He forgot all about the pyramid sitting on the bookshelf. Nine days later he peered into the pyramid in passing and saw that the candle had turned into a crippled wax finger. The deformation of the candle could not have been caused by the moderate fall temperatures, and none of the other candles in the room were affected in any way.

Mrs. Elka, a twenty-six-year-old amateur painter from Wuppertal, Germany, was similarly astounded. She enjoys creating miniature oil paintings a mere five centimeters wide. After completing one of her pictures, she positioned it on a delicate pedestal inside a twenty-eight-centimeter-high glass pyramid. She was not interested in an experiment. Rather, she felt that the miniature picture, which showed a small house, a cat, and a full moon, looked particularly pretty behind the triangular glass walls of the pyramid. One week later, Mrs. Elka thought that the painting had changed. Three weeks later, the moon had dripped from the sky, the paint of the dark brown wooden roof had become crusty, the dark blue sky had a brilliant shine, and the end of the cat's tail had disappeared. Great special effect! I recommended that Mrs. Elka try to sell her future creations as genuine pyramid paintings.

Another pyramid experiment, conducted by Mr. and Mrs. Burgmüller in Hamburg, Germany, involved the use of ordinary honey. The Burgmüllers, who live on the eighth floor of a high-rise building, bought a small Plexiglas pyramid, only fourteen and a half centimeters high. One day, after breakfast, Mr. Burgmüller measured two tablespoons of honey into a small bowl, which he placed on a podium inside the pyramid. Twenty-four days later, the honey had hardened into a solid lump which felt like stiff wax. Then, inadvertently, Mrs. Burgmüller moved the pyramid out of its north-south direction when she was cleaning the living room. Presto! Six days later, the honey was running from the bowl again. Maybe this could explain the tears of Saint Januarius in the cathedral of Napoli, which magically begin to flow every year.

The above, rather haphazard results have been confirmed by more scientific experiments. Some quiet and friendly individuals scrupulously document their experiments and even weigh their objects on letter scales. Gerhard Leiner of Graz, Austria, constructed a model pyramid from particle board, four and a half millimeters thick. He began his experiments on March 19, 1983, at twelve-thirty. Inside the north-south-oriented pyramid was a seven-day-old chicken egg weighing 60.2 grams. A second egg was kept outside the pyramid. The room in which the experiment took place had an average temperature of 19 degrees Celsius.

On October 4—two hundred days later—the egg inside the pyramid had lost 58.8 percent of its weight, but the yolk was yellow, the smell was normal, and the egg was edible. The control egg outside the pyramid, on the other hand, was reeking to high heavens (or to the room ceiling, anyway). Leiner conducted other long-term experiments which confirmed the earlier results. However, none of the eggs have hatched so far.

Other members of the Ancient Astronaut Society experimented with apple slices, radishes, flower seeds, tobacco, orange juice, cucumber and tomato plants, even strawberries. The experimenters agreed unanimously that all the fruits inside the pyramid acquired a stronger taste. Vegetables under a foil-covered greenhouse pyramid grew faster than control plants; cucumbers and tomatoes were firmer, denser, and had a much more concentrated aroma that any control group of vegetables.

Magic? Hocus-pocus? Fraud, or an overactive imagination? Imagination is the only weapon in the war against reality, it is true—but it was not a factor in these experiments. The objects in the experiments changed measurably and visibly. The scientific method requires that results be reproducible at any given time. Well, they are, although no one knows what is really happening or why.

Some friends of mine gave me a glass pyramid as a gift. It sat around the porch, a winter garden of sorts, for several weeks. One night I had some Bordeaux wine that had not been properly aged. I admit: I enjoy having a glass of Bordeaux occasionally. In time, the palate, the tongue, and the stomach can tell whether a wine flows smoothly, whether it is of good quality, whether it soothes the stomach and spreads like divine nectar through one's body. This Bordeaux was wild, raw, and sour. It had not matured. While I was pouring the wine into a vinegar bottle, I was possessed by the pyramid spirit, which made me do something very strange. I took another Bordeaux of the same brand and placed it inside my glass pyramid—and forgot about it. Autumn and winter came and went. In the spring, I, the emancipated husband, helped my wife straighten up the porch. There was the wine bottle!

The color of the Bordeaux had become darker. It tasted full-bodied, silky, nonacidic—like a seven-year-old grand cru classé. (The expert knows what this means.) I set up a taste test with a second bottle of the same vintage, which had been sitting in the cellar. The difference was striking. Regular guests to my house can attest to the fact that ever since that day I have kept a bottle of Bordeaux under my pyramid. For several occasions!

Suggestions for Possibilities

Pyramid numbers and pyramid forces—they exist, yet no university attempts to understand them. Shouldn't immunologists and hygienists be curious to find out why certain bacteria, viruses, and fungi perish inside the pyramid, while others do not. Does the pyramid shape alter poisons that are difficult to dispose of? Does it harden alloys or welding joints? Can pyramids improve the effectiveness of crude oil or other naturally occurring chemicals, enhance the flavor of spices, or even clean pool water without clorine? Can pyramids be used as water treatment plants? As fresh water storage tanks? Would it be possible to age wine by the barrel inside a pyramid, preserve fresh vegetables, flowers, and fruits? As a world traveler, I know firsthand how quickly medicines spoil in third-world countries because refrigerators either are not available or are not functioning. Why hasn't it occurred to any of the huge chemical concerns to try pyramid packaging?

I am merely jotting down some spontaneous questions. Thoughts have effects. Maybe one of my ideas will stimulate some alert mind. It would be a shame to let the pyramid powers go to waste just because they carry the slight stigma of obscurity. After all, the most bothersome fact about these pyramid phenomena is their proven existence. How often have spontaneous utterances produced great things? So, I am letting my thoughts flow freely that they may inspire greater glory.

Do you feel drained? Tired? Depressed? Lie down inside a pyramid for a couple of hours so that your head rests in the lower third of the pyramid. You will notice with astonishment how the neurons in your burnt-out brain are starting to rejuvenate. However, be careful not to prolong the effect indefinitely—the pyramid effect is known to shrink water heads.

Are you unable to solve a particularly sticky problem? Are you searching for a truly exceptional idea? An overwhelming inspiration? The magic of the pyramid can do the trick. I have experienced it myself, with great amazement.

For decades, radio astronomers have been attempting to make contact with extraterrestrial life-forms in outer space. So far they have been unsuccessful, because their means and the wavelengths they employed have been limited. Radio astronomy is based on electromagnetic waves—what else?—because radio waves travel three hundred thousand kilometers per second and thus are the fastest means of communication—fast in earthly terms, but not fast enough in universal terms. Talking to extraterrestrials in a solar system twenty light-years from here would be very boring. It would take at least forty years before the answers to our burning questions would reach our antennas. Isn't there anything faster than radio or light waves? Is the pyramid shape a transmitter to the universe, our ear to the extraterrestrials? Do the magnetic forces of the earth enhance our thoughts inside a correctly positioned pyramid? Do people who pray send out thought patterns with praises and wishes to an eternal creator via a resonance floor in a church or temple? Can the powers of the pyramid transform human thoughts into impulses that travel faster than the speed of light? Are extraterrestrial telepathists in space pyramids waiting out there for our messages?

Would you like to be a time traveler? Would you like to be carried away by the wave of time into the past or future? Would you like to progress, just once, to a different dimension and meet alien beings? Paul Brunton, a historian, spent a night inside the Great Pyramid and reports that wondrous things happened there. He describes the climax of that night, when gigantic monsters, horrible images from the underworld, figures with grotesque, crazy, terrifying, devilish looks, surrounded him and filled him with extreme disgust. In the span of a few minutes, he experienced something that he says is etched into his memory like a photograph.[38]

In the course of the night, Paul Brunton had contact with the high priests of an ancient Egyptian cult and was transformed into a spiritual being and led to the "teaching hall." He was told that the pyramid stores the memory of lost tribes and the covenant which the creator made with the first great prophet. Brunton even maintains that the spirits led him into a hall deep below the pyramid.

Does or did the Great Pyramid serve as a storage place for documents of ancient cultures? Are there any chambers and passageways that have not been explored? At what time in human history was this "time capsule" supposedly created? Does this hall Paul Brunton described, deep below the pyramid, actually exist?

It does—I have been there.

✧ 4 ✧

The Eyes of the Sphinx

"I simply picked a bunch of flowers and added nothing but the thread that binds them."

MICHEL DE MONTAIGNE,
FRENCH WRITER (1533–1592)

It is the beginning of December 1988. The plateau of Giza is deserted. No tourist buses, no honking, no jostling, no camels, horses, or pesky vendors, no lines in front of the entrance to the Great Pyramid. The streets and paths between the ancient buildings are swept clean like the famous Bahnhofstrasse in Zurich. Schoolchildren are playing outside, irreverently bouncing their balls off the pyramid stones. Two serious-looking guards are blocking the entrance to Khufu's world wonder. They let no one pass, not even individual tourists who may appear at their gate.

But no one is coming. What is happening in Giza? Are foreigners no longer welcome? A friendly inspector explains that the Grand Gallery is being restored. All travel agents and hotels have been informed, and so the tourists are not even brought out to Giza. Egypt has an endless number of splendid temples to offer instead, and the pyramids at Saqqâra amply compensate visitors for missing out on Giza.

My companion, Rudolf Eckhardt, a superb amateur photographer, and I introduced ourselves to the young inspector and politely asked him to make an exception in our case. We told him, truthfully, that we would appreciate the opportunity to take some pictures inside the Great Pyramid without the usual hustle and bustle of tourists. The man invited us into the Egyptologist's office. Several students and inspectors were sitting around on an old couch and some chairs. They patiently listened to my plea and took turns scrutinizing my identification papers, stealing furtive glances at our camera equipment.

"Video? Film?" asked the chief of the group.

"No," I answered, smiling reassuringly. "Just pictures!"

They offered us sweet black tea, and I handed out some chocolate from Switzerland. We talked shop for a while. I was glad that I had read so much about Egypt in the past few years. Then the chief asked one of the students to accompany us. We marched over to the Great Pyramid. The student asked us kindly if we needed any assistance.

"No," I replied. "We are familiar with all the pertinent literature about the pyramid. We would just like to take some pictures without being interrupted by tourists."

In front of the steps to the pyramid entrance, our student companion met two colleagues. They exchanged greetings. I told the student that he could wait there, that we would take the pictures and then return to this spot. He nodded and called out some orders to the guards at the entrance. "Salaam." They bowed modestly and let us in.

A Vault in the Rock

Immediately upon entering, we noticed that the path to the ascending corridor was not the one along which tourists are generally admitted. A gently winding tunnel, broken out of the stone blocks, led us inside. As with any other visit, I was forced to stoop as I pulled myself up to the Grand Gallery with the help of wooden handles secured to the walls. What a sight! The pyramid had not seen anything like this in forty-five hundred years. The whole gallery was stuffed with metal scaffolding and boards. We could not see any of the details which we had come to photograph. So we were happy to observe that at least the gate to the Queen's Chamber, which is usually shut, was standing open. But it offered the same sight: scaffolds, boards, and ladders. We turned around and walked back to the so-called three-way crossing, the point at which the ascending and descending corridors intersect with the entrance tunnel. Lightbulbs spread a gentle, constant light.

The gate leading to the corridor which descends deep below the pyramid was open as well. I peered down into the seemingly endless shaft. Gradually, the lights along the walls grew fainter as they disappeared into the gaping abyss. I had read about the subterranean sepulcher at the bottom of the pyramid. The inspectors rarely grant one permission to visit the vault; they say that the climb is too strenuous and too dangerous. Now we were standing in front of the entrance to the tunnel, and there was not a guard in sight. On the contrary—two guards were outside the pyramid entrance barring anyone else from entering the pyramid. We called out a few times, "Hello. Is anybody there?" But only our own voices echoed from the walls. We were alone inside the pyramid.

The shaft measured 1.2 by 1.06 meters—not high enough to allow one to stand upright, yet high enough that crawling wasn't necessary. I swung one of my cameras onto my chest, the other onto my back, ducked my head and shoulders, and waddled down the corridor, hunched and flat-footed like a duck. Rudolf, who was carrying even more equipment, followed behind. I repeatedly shone my flashlight on the smoothly polished white Tura limestone walls. What superior craftsmanship! The joints between the stones are barely noticeable and follow the angle of descent rather than running horizontally. The angle of inclination is 26 degrees, 31 minutes, 23 seconds. We were panting quietly. After about forty meters we took a break. My hair was caked to my forehead. On we waddled, duck style. After about sixty-five meters, we came upon a recess in the wall, on the right-hand side. Fresh air was blowing through an ancient pipe, thousands of years old. On and on we went. Would this corridor never end? My thighs were throbbing. My tendons are unaccustomed to such exercise. Eighty meters . . . ninety meters . . . we could not see any more lights below. We both knew that the corridor ends in a vault, but neither of us was aware of the length of the passageway. After another twenty-eight meters I felt raw dirt under my shoes.

The air was sticky, warm. We were finally able to stand. A spotlight was lying on the ground, gutted wires hanging frazzled from its torn cable. With trembling hands, in the beam of my flashlight, Rudolf tied the ends of the cable back together, anxious to avoid short-circuiting the lamp or being electrocuted himself. The light came on.

The vault in which we were standing was about thirty-five meters below the pyramid's foundation. Ancient Arabian sources tell us that the first person to enter the vault was Caliph Abdullah al-Mamun, the son of the famous Harun al Rashid, a character in the *Arabian Nights*. Al-Mamun ascended to the throne of Bagdad in 813 A.D. and also ruled Egypt from 820 until his death in 827. The young caliph was considered an intelligent man who supported the sciences and was aiming to improve the status of Arabia in the world. Ancient documents reported that thirty secret treasure vaults were located under the Great Pyramid. The vaults were said to contain the divine ancestors' precise land and sky maps. No wonder al-Mamun was interested in these treasures. And since he was the ruler of Egypt, no one could blame him. Besides, the Moslem priests considered the pyramids heathen structures and had no objections to their desecration.

How to Crack a Pyramid

Al-Mamun formed a crew of craftsmen, workers, and architects to find a way into the pyramid. But no number of crowbars or chisels could move even a single stone from the pyramid wall. Finally, al-Mamun's men remembered an ancient warfare technique used to break up walls. They lit a big fire directly in front of one of the pyramid blocks and fanned the fire until the stone was searingly hot. Then they poured vinegar onto the rock. It cracked and finally could be smashed with battering rams. In this way, al-Mamun's crew created the entrance which is still used today by tourists.

With great effort, the men forced their way about thirty meters into the pyramid before the air became too thin, sticky, and poisonous, because the fire and the torches were consuming all the oxygen. The men were discouraged, ready to report to their master that they had failed. Suddenly, they stopped. Inside the pyramid they could hear a dark rumbling sound, then a loud crash. They had to be close to a passageway. Somewhere inside the pyramid, a stone had tumbled down.

Encouraged, the men continued their drilling, hammering, hoisting, and chiseling and finally reached the descending corridor which Rudolf and I had just waddled down. Initially, al-Mamun's men were not interested in the descending portion of the passageway. Instead, they crawled upward until they reached the actual secret entrance to the Great Pyramid, situated sixteen and a half meters above ground level, ten layers of rock higher than the hole al-Mamun's men had opened up. After some pep talks and prayers to Allah, the men followed the dark corridor down to the spacious vault in which Rudolf and I were now standing.

The spotlight fell on the ceiling, which was hewn out of the solid rock. It wandered over to the walls and illuminated two monolithic pedestals of gigantic proportions. Two curious, unfinished bulges protruded from the monstrous rocks. On the floor behind us we discovered a rough-edged shaft, about four meters deep, surrounded by an iron guardrail. In the southeast wall to its left was an opening identical in size to the corridor through which we had just passed. Very adept at duck waddling by now, we crawled into the passageway, anxious to discover what other chambers the tunnel might lead to. After about fifteen meters, the corridor ended abruptly. A dead end down here? What for?

The underground room that has been chiseled out of the rock below the pyramid measures 14.02 meters in an east-west direction and 8.25 meters from north to south. Those are respectable dimensions. Modern archaeologists call it an "unfinished burial vault"[1] and thus plunge us into a hurricane of absurdities.

Contradictions

The pseudo-sepulcher is supposed to be "unfinished"? Let us think this through in detail. It is unlikely that the cave was created after the pyramid had already been erected. How would they have disposed of the waste materials? It is safe to assume, then, that the subterranean structures were completed before the top of the pyramid was added on. How did the masons get down to thirty-five meters below ground? Through digging and chiseling, of course. Like a mole, the first worker in line must have broken fragments out of the rock with soft copper and iron chisels and moved them down the line to his colleagues behind him, who transported the rubble up to the entrance. The deeper the tunnel descended underground, the darker it got. Right? They needed torches, wax, oil lamps. There goes the last breath of oxygen.

Obviously, this scenario is out of the question. The

builders must have used air shafts, as we know them from later mining operations. But where are they? Today, we know of one shaft running across the descending corridor, and we are told that it was created by grave robbers. Never mind. Let us presume that the human moles, some way or other, reached the point at which they planned to build the underground sepulcher. Now they continued as suggested above: "Let's get out the hammers and chisels, friends! Who needs light and air down here?" Maybe the crews worked in the dark, with radar, x-ray, or albino vision, and did not give a hoot if a piece of rock crushed someone's head, pinched some fingers, or squashed some feet now and then. The rubble was hauled up to the surface with sleds. Air was probably filtered into the dusty vault through hoses made of animal intestines.

My sarcastic description is intended to demonstrate how it certainly was *not* done. Undoubtedly, there must be air shafts leading to this subterranean chamber under the pyramid. Archaeologists, light your lanterns and knock on all the walls and ceilings! Maybe you will be fortunate enough to find one of the treasure vaults described in ancient tales.

After the cave was half done, the merry workers must have dug a fifteen-meter-long, dead-end tunnel in the southwest corner, just for the fun of it, and lined it with nicely polished stones. To top it off, they dug a hole in the ground, abandoned the cavernous vault, and—picture this!—began to line the corridor, which they had broken so painfully out of the rock, with massive, smoothly polished Tura stones. Over one hundred meters, straight as an arrow, pointing upward without the slightest angle deviation. And why did they go through all this trouble, labor, and blood-spitting effort in a narrow dungeon? To create an unfinished hole in the rock, thirty-five meters below ground, which was never going to store anything anyway.

Some people live so cautiously that they die practically brand-new; others use their brains exclusively for reading, never for thinking. I am told that a different architect or builder took over during the construction of the pyramid and that the plans were changed on short notice. I beg your pardon? As long as stones were hammered out of the rock in the "unfinished burial vault" below and as long as rubble needed to be transported up to the surface, the descending one-hundred-meter shaft could not be lined with polished Tura stones. Even with no more than ten meters of the Tura stones in place, it would have been impossible to dispose of the waste materials from the cave below. There simply was no room—I know, because I crawled down there. Moreover, the rubble would have scratched the perfectly polished walls. There are no scratches, and neither are there any indications that wheels or dragging devices were used. If we follow the archaeologists' interpretation that the cavernous hole is an "unfinished burial vault," a cave which suddenly lost its purpose and became superfluous in the plans of a new builder, there would be no reason to line the 118-meter-long access corridor to the useless sepulcher with polished Tura monoliths. After all, the polished stones could not be installed until after excavation on the subterranean chamber was complete. A royal entrance to an unfinished dirt hole below the pyramid? A dead-end tunnel issuing from the same chamber? What's wrong with this picture?

I see three possibilities:

1. The chamber is continued somehow, behind a monolith maybe.
2. The vault has already been plundered.
3. Someone was put to rest in the sepulcher, maybe in a hibernating state. This unknown person did not care about earthly names, inscriptions, or praises, or about a chamber filled with monoliths. He was interested in the preservation of his body, and nothing else. The body had to survive a certain amount of time without decay. There was no need for decorations and gewgaws.

Perhaps all three of these possibilities are somehow intertwined.

What did al-Mamun's brave intruders find inside the "unfinished burial vault"? What did they, the first ones to enter it in several thousand years, find inside the Great Pyramid?

The Arabs' Exciting Discoveries

Nobody knows anything specific. There are no written inventories, or if there were, they have been lost. In the fourteenth century, libraries in Cairo still had old Arabic and Coptic manuscripts and fragments, which the geographer and historian Taki ad-Din Ahmad ben 'Ali ben Abd al-Kadir ben Muhammad al-Maqrizi compiled in his opus *Khitat*. The quotes from this book are worth savoring. While some passages may remind us of the flowery Arabic expressions of the *Arabian Nights*, they also contain a substantial amount of names, dates, traditions, and other amazing information. The *Khitat* states that the three great pyramids were built under a favorable constellation that had been agreed upon.[2]

> After that, he [the builder] had thirty treasure vaults made from colored granite in the western pyramid; they were filled with rich treasures, with tools and stelae made from precious stones, with tools made from excellent iron, like weapons that will not rust, with glass that can be folded without breaking, with strange talismans, with the various kinds of simple and combined healing potions and with deadly venoms.
>
> In the eastern pyramid, he created representations of the various constellations and the planets, as well as pictures of the things his ancestors had created; added to that were incense sacrificed to the stars and books about the stars. One also finds there the fixed stars and that which happens in their eras from time to time. . . . Finally, he had the corpses of the prophets brought to the colored pyramid in coffins made from black granite; next to each prophet was a book which described his marvelous skills, his life, and his works that he had accomplished in his time. . . . And there was no science that he did not have written down and recorded. In addition, he had the treasures of the constellations brought there, which had been offered to them as gifts, as wells as the prophets treasures, and they made up a huge and innumerable amount.

Furthermore, we are told that the king had an idol placed under each pyramid to fight off possible intruders with different weapons. One of these guards was "standing upright and had some kind of spear with him. A snake was curled around his body which would attack anyone approaching the guard." Another guard is described as having wide-open, flashing eyes and as sitting on a throne, also carrying a spear. Those who saw him were unable to move and stood there, immobilized, until they died. The third pyramid was protected by a guard who grabbed intruders and pulled them close until they stuck to him and could not get away, and finally died. When the builder of the pyramids died, he was buried inside one of the pyramids.

Arabic traditions hold that all three pyramids contained treasures and books with unbelievable contents. Did al-Mamun plunder the treasure vaults? Did he find mummies in sarcophaguses?

> Al-Mamun opened the Great Pyramid. I entered the inside and saw a large, vaulted chamber, with a rectangular base, but round at the top. In its center was a rectangular well, ten yards deep. If one descends down the well, one finds a door on all four sides which leads to a large room in which corpses are laid out, sons of Adam. . . . It is reported that people climbed up there in al-Mamun's time and reached a vaulted chamber of small size, in which stood the statue of a man, made from green stone, a kind of malachite. It was taken before al-Mamun and discovered that it was closed off with a lid. When it was opened, one saw inside the corpse of a man wearing a golden coat of armor embellished with all kinds of precious stones. On his chest lay the blade of a sword, without handle, and next to his head a red hyacinth stone the size of a chicken egg, which glowed like the flame of a fire. Al-Mamun took the stone. But I saw the statue out of which the corpse was taken next to the gates of the royal palace in Misr [Cairo] in the year 511.
>
> . . . they now entered the middle chamber and found three biers made of transparent, glowing stones; on them were three corpses; each was covered with three robes and

> had next to his head a book written in an unknown language. . . . Al-Mamun ordered that everything that had been found in the vaults be removed; then, at his command, the figures on the columns were pulled down again, and the gates closed like before.

We are tempted to discard this account as being a little oriental, too sappy to be true. But what gives us the right to judge solely from our present viewpoint and to regard ancient reports as unreliable? Were any of us there? Did any of us know the chroniclers, who in their time were honorable and well-respected men? We see ourselves as a society with electronic mass communication, the best-informed society of all time, but every piece of information which scientists, students, journalists, the media, and ordinary people receive has already been screened, filtered, and partly distorted. The opinion we form about an issue is often no more than someone else's regurgitated opinion, and that opinion, in turn, was formed on the basis of biased information. Generalizations like "Arabian chroniclers are giving us fairy tales," or "We know everything there is to know about the pyramids," or "The proven, scientific opinion . . ." are phrases which expose our lack of knowledge. We have become myopic, probably because the flood of information with which we are inundated forces us to screen out certain ideas. All too often we *think* we know something.

The Arabian historians tell us that al-Mamun found the corpse of a man wearing a strange coat of armor decorated with precious stones. Is it a fairy tale? We know such armor from the Old Testament. The second book of Moses, Exodus, chapter 28, details the garments which Aaron (Moses' brother) and the Levite priests were told to wear: specifically, among other things, a breastplate with twelve different precious stones.

New Corridors and Chambers

There are supposed to be statues, sarcophaguses, and books with scientific information in the three great pyramids. Derivations of an overactive imagination? Doesn't "science" know all there is to know about the pyramids already? Gullible minds may think so.

The radiation experiment conducted on the Khafre pyramid in late 1968 and early 1969 by Nobel Prize–winning physicist Dr. Luis Alvarez is well-known. Alvarez and his team based their studies on the fact that cosmic rays bombard our planet twenty-four hours a day and that they lose some of their energy as they penetrate solid matter, such as rock. On the average, a square meter of soil is flooded with ten thousand protons per second. The most highly energized of these cosmic particles can penetrate even the thickest layers of rock; others can pass through the entire planet. The number of elementary particles entering a rock formation can be measured with instruments. If the rock contains any hollow spaces, the protons slow down less in passing through the hollow space; thus the flow of protons is greater than in solid rock.

A spark chamber was installed inside the Khafre pyramid, so that the rays of the cosmic particles could be recorded on magnetic tape. These tapes were then analyzed by an IBM computer, taking into account the shape of the pyramid, its size, and its angle of inclination.

By the end of 1968, the orbits of over two and a half

million cosmic rays had been registered. The computer analysis showed the shape of the pyramid correctly. The scientists knew that the experiment was reasonable and the instruments were working properly.

Then came the big surprise that left the scientists aghast. The oscilloscopes recorded a chaotic pattern. There was no discernible order anymore. If the same magnetic tape were run through the computer twice, it came up with different data and charts each time. It was incredibly frustrating. The experiment, which was very expensive and was funded by various American institutions, IBM, and Ain-Shams University of Cairo, yielded no measurable results. Dr. Amr Gohed explained to journalists that the findings were "scientifically impossible," and he added that either the structure of the pyramid is a mess or there is some kind of mystery which we cannot explain, whether one calls it the occult, the curse of the pharaohs, or magic.[3]

Other scientists have searched for rooms inside the pyramid since then, with new devices and new methods. And they have been successful. In the summer of 1986, two French architects, Jean-Patrice Dormion and Gilles Goidin, located hollow spaces inside the Great Pyramid with the help of electronic detectors. In cooperation with the Egyptian Department of Antiquities, they sent microprobes through two and a half meters of rock. Below the corridor leading to the Queen's Chamber, the French scientists found another chamber, about three meters wide and five and a half meters high, filled with crystalline silica. Another hollow space was located behind the northwest wall of the Queen's Chamber. So far, no entrances to these chambers have been discovered. So what do we know? How can we dismiss Arabian traditions as fairy tales?

Alarmed by the success of the French team, the Japanese researchers at Waseda University in Tokyo stepped up their efforts. They had already tested a kind of radar device which could practically x-ray various rock formations—granite, limestone, sandstone. The top-notch crew from Waseda University arrived on January 22, 1987, in Cairo. It included a professor of Egyptology, a professor of architecture, a doctor of geophysics, and various electronics engineers. The leader of the team was Professor Sakuji Yoshimura, who established close professional cooperation with Dr. Ahamed Kadry, the director of the Egyptian Department of Antiquities.

The Japanese, always brilliant in the electronic sector and equipped with superb mobile instruments and computers, x-rayed the corridor leading to the Queen's Chamber, the Queen's Chamber itself, the King's Chamber above, the entire area south of the Great Pyramid, and finally the sphinx and its surroundings. What did they find? I will make it brief: The Japanese researchers discovered unmistakable evidence that there is an entire labyrinth (!) of corridors and chambers lurking inside the Great Pyramid.

The colorfully illustrated scientific report of the Waseda University team[4] presents sixty pages of actual measurements taken at the individual sectors, and they are full of white bars—corridors, shafts, and empty spaces inside the pyramid. Southwest of the King's Chamber and southwest of the main axis of the Grand Gallery, fairly large vaults were located. A corridor originates at the northwest wall of the Queen's Chamber, and a forty-two-meter-long pit was found south of the Great Pyramid that seems to extend below the pyramid. The electronic devices of the Japanese also indicated the existence of a second sun bark in the rock below the pyramid, and this finding has been confirmed since then.

Now what? What other surprises are in store for us? How do those all-knowing scientists who had dismissed the possibility of undiscovered chambers in the pyramid feel now? At present, no one knows what is inside those electronically detected vaults and corridors—or whether they have been plundered already.

No one? As I told you before, the Grand Gallery and the Queen's Chamber were stuffed with scaffolding and boards in December 1988. There was not a construction worker in sight. My question is: Did additional electronic tests and drillings take place in the dark of the night? Were microprobes with glass fibers inserted through the pyramid stones to photograph the inside of the pyramid? I could understand such procedures; it is impossible to do any kind of scientific study amidst the tourist buzz. On the other hand: Wouldn't it be detrimental to the reputation of the field of Egyptology if researchers opened these chambers, hidden for thousands of years, like thieves in the night, without informing the public? Who would believe, after the fact, that the artifacts presented—or not presented—were all that was found?

The Khufu Hoax

Maybe we are in for a surprise of a different sort in the Great Pyramid, one that would be particularly painful to Egyptologists—namely, the fact that Khufu was not the builder of the Great Pyramid. Whenever I ask one of the experts who the builder of the Great Pyramid was, I get the same immediate, stereotypical answer: Khufu. "No doubt?"—"No doubt." Pharaoh Khufu is a proven scientific fact. It is inappropriate to question it. And that's that! But if we dig a little deeper below the surface, this "proven scientific fact" is in for a rude awakening.

What evidence bestowed the halo of the pyramid builder on Pharaoh Khufu? Where does this absolute certainty come from that no one but Khufu could have built this most impressive edifice? Remember that there are no texts in the Great Pyramid praising or glorifying the builder. It is anonymous vanity.

Under closer scrutiny, we find that there are only two indications that point to Khufu, but scientific literature has blown these two out of proportion. Herodotus wrote that the pyramid was commissioned by Cheops. "Cheops" is Greek for the Egyptian "Khufu." Diodorus Siculus lists "Chemmis" as the builder of the pyramid, but Gaius Plinius Secundus, who enumerates the names of the historians who described the pyramid before him, notes wryly that none of them really knew who the builders were. In this particular case, archaeologists rely exclusively on Herodotus—while discounting his statements everywhere else.

The second piece of evidence pointing to Cheops/ Khufu is an inscription in one of the corbeled vaults above the King's Chamber. Whoa! Didn't I claim repeatedly that there are no inscriptions inside the Great Pyramid?

It is a mystery, or actually a crime, perpetrated by an imposter. The case was not solved by Sherlock Holmes, but by Zecharia Sitchin, a specialist in old oriental languages.

On December 29, 1835, Col. Howard Vyse, a British officer of the Guards, arrived in Egypt. Vyse was a sardonic gentleman, a grandson of the Earl of Stafford, on the one hand disciplined to the bone, on the other the black sheep of the family who had to work extra hard to prove himself. Vyse was thrilled and fascinated by the mystery of the pyramids. He immediately hooked up with the Italian captain Giovanni Battista Caviglio, who had been digging in Giza for quite some time. But over the course of the next few months the two men had a falling-out, and they parted company on February 13, 1837. Vyse, the Briton, who had received a license to excavate from the British consul, chased the Italian off.

Seventy-two years prior to Howard Vyse, the British diplomat Nathaniel Davison had discovered a hole in the ceiling at the end of the Grand Gallery. On July 8, 1765, he crawled inside. Davison progressed to the lowest one of the corbeled vaults above the King's Chamber. Howard Vyse, of course, was aware of Davison's discovery. He noted in his diary that he suspected the existence of a sepulcher hidden above the "Davison Chamber." Vyse was determined to become famous. He wanted his name to go down in history; he felt he owed that much to his family. On January 27, 1837, he wrote in his diary that he had to discover something before returning to England. Vyse and his chief engineer, John S. Perring, procured some gunpower and blew a tunnel into the rock above the Davison Chamber. Indeed, on March 30, April 27, May 6, and May 27, 1837, Vyse and Perring discovered four additional chambers above the Davison Chamber, which they named the Wellington, Nelson, Arbuthnoth, and Campbell Chambers, in that order. In the upper two rooms, Vyse noticed some signs on the monoliths which apparently had been painted with red paint. It was known that builders often marked individual monoliths from the quarries of the Wadi Maghara mountains with paint so that there would be no mix-up and they would reach their proper destinations. One of the paint marks displayed the pharaoh name "Khufu." It was the proof that was needed: The painted monolith was shipped to Khufu/Cheops. The sensational news spread around the globe. Howard Vyse had accomplished his goal.

No one seemed to mind that the Khufu insignia, which should have been found all over the place, considering that more than two million stones were used in the Great Pyramid alone, were not found anywhere else.

In a fascinating book[5] and in two articles in *Ancient Skies*,[6] the American Orient expert Zecharia Sitchin exposed Vyse as a fraud. The evidence against Howard Vyse is such an ingenious, criminological masterpiece that it is difficult to understand how Egyptologists can continue to defend their erroneous "proven scientific fact."

Zecharia Sitchin unmasks the Vyse-Perring hoax with the help of dates, quotes, and excerpts from Vyse's diary, and in particular because of an orthographic mistake committed by the forger. Shortly after the "Khufu" paint marks were discovered, some experts voiced doubts about its authenticity, but their words of caution were drowned out by the prevailing euphoria. Samuel Birch, an Egyptologist and expert in hieroglyphics, cautioned in 1837,

"Although [the symbol] is not very legible because it is written in semi-hieratic or linear-hieroglyphic letters . . ." and, a littler later, "The meaning . . . is not entirely clear . . . very difficult to interpret. . . ."[7]

What caused such confusion for the hieroglyphics expert? The painted marks used symbols which did not even exist in Khufu's time. Over the centuries, the picture symbols of ancient Egypt turned into a "hieratic language"—long after Khufu. Even Richard Lepsius, the (presumed) discoverer of the Labyrinth, wondered about the brush-painted red symbols because they resembled the hieratic language.

How did the symbols end up inside the Great Pyramid? Did anyone enter the pyramid many centuries after its completion to paint the symbols on the monoliths? No way. The corbeled vaults were completely inaccessible. Vyse had been forced to use gunpowder.

Vyse, a military man, not an Egyptologist, knew only one standard book about hieroglyphs: the textbook *Materia hieroglyphica* by John Gardner Wilkinson, published in 1828. It was discovered later that the name "Khufu" is misspelled in Wilkinson's textbook. The consonant "Kh" was mistakenly shown as the sun symbol "Re." The impostors Vyse and Perring not only used a language which did not develop until many centuries after Khufu, they also copied the orthographic mistake from Wilkinson's textbook. Didn't anyone realize that the red paint was fresh? Zecharia Sitchin[8] commented that this question was answered at the time by one of the principals, namely Perring, in his own work about the pyramids of Giza. Perring wrote that the paint that was used for the ancient Egyptian inscriptions was a compound of red ocher, which the Arabs called *moghrah* and which is still in use. The drawings were preserved so well on the stones that one could not possibly tell if they had been done the day before or three thousand years earlier.

I asked several Egyptologists about the discoveries Zecharia Sitchin had made. None of them were familiar with his analysis. They assume that they know the truth and comfort themselves with the assurance that Howard Vyse was an honorable archaeologist. Vyse was not an archaeologist. He may have been honorable in certain ways—but he also was a glory hunter.

Honor is a dubious quality, even in archaeology. The British archaeologist Howard Carter gained world fame on November 4, 1922, when he discovered the tomb of Tutankhamen. No one dared question his side of the story; his reputation was beyond reproach. Carter maintained that, unfortunately, the anterooms of the actual tomb had already been opened by grave robbers. Today we know that Carter was lying through his teeth. He was the one who entered Tutankhamen's vault *before* the official opening of the tomb, turned everything upside down, and stole a number of valuable objects so that he did not have to share them fifty-fifty with the Egyptian government, as his contract demanded. This crime was exposed by Dr. Rolf Kraus, an archaeologist at the Egyptian Museum in Berlin.[9] Yet neither the experts nor the public reacted to Kraus' discovery.

Who Was the Builder?

There is not a shred of convincing proof that Khufu was indeed the builder of the Great Pyramid. This does not invalidate the possibility that he may have built the pyramid, but there is more evidence against him than for him. There are no hieroglyphs, pyramid texts, statues, busts, walls full of praises. A single, tiny, five-centimeter-high figurine made of ivory and displayed in the Antiquities Museum is supposed to represent Khufu. On the other hand, there is evidence, solid as a rock, *against* Khufu; yet the experts ignore it.

In 1850, a stele was found among the ruins of the Isis temple which is now exhibited in the Egyptian Antiquities Museum in Cairo. The Isis temple was located directly adjacent to the Great Pyramid. The inscription on the stele states that Khufu founded the house of Isis, the goddess of the pyramid, next to the house of the sphinx. Since Isis is mentioned as the goddess of the pyramid, it is clear that the Great Pyramid was already in existence at the time of Khufu. Moreover, the sphinx is said to have existed as well, although archaeologists claim that it was built by Khufu's successor, Khafre. Why do scientists ignore this explosive, set-in-stone message? The stele was found in 1850. Remember: Thirteen years earlier, it had been determined that Khufu was the builder of the Great Pyramid, thanks to Howard Vyse's fraudulent discoveries. The stele did not fit the scientists' models. Archaeologists declared it a fake and claimed that it was created after Khufu's death to support the ideas of the local priests.

All of these elements justify my question: If Khufu was not the builder of the world wonder of Giza, then who was? Egyptologists have complete chronological accounts of all the pharaohs after Khufu. There is no room for an additional pharaoh *after* Khufu. If it could not have been a pharaoh after Khufu, then it must have been one *before* Khufu. The very thought is unbearable to archaeologists. It would completely upset their treasured chronology of construction events. Can Arabian historians help us out? What does al-Maqrizi have to say?[10]

> The largest pyramids are those which are across from Misr to the present day. People cannot agree on the time of their construction, the name of the builder, or the reason for their construction and have expressed various opinions, most of which are wrong. I now want to relate that which satisfies and suffices, if God, the Almighty, allows me to.
>
> The teacher Ibrahim Ben Wasif Sah al-Katib states in his *News of Egypt and its Wonders*, in the place where he tells of Saurid, the son of Sahluk, the son of Sirbak, the son of Tumidun, the son of Tadrasan, the son of Husal, one of the kings of Egypt prior to the Great Flood, who resided in the city of Amsus, which is discussed in the part of the book which deals with Egypt's cities. He was the builder of the two great pyramids near Misr. . . . The reason for the construction of the two pyramids was that three hundred years before the Great Flood, Saurid had the following dream: The earth turned upside down with all its people, the people fled in a blind rush, and the stars fell down. . . .

Considering the precise list of names, I find it difficult to discard the text as a fairy tale or myth. According to the description, a king named Saurid had a dream three hundred years *before* the Great Flood which eventually led to the construction of the pyramid. His advisors and prophets were also plagued by horrible dreams. They predicted the end of civilization. "The sky opened up, a bright light emerged . . . and men descended from heaven who carried iron clubs in their hands which they used to bludgeon people."

Older Than the Flood

The king asked his wise men if people would be able to live again in Egypt after the Flood. Their answer was affirmative. He decided to build the pyramids so that all the human knowledge of the time would be preserved. An excellent reason. At the top of the pyramid, the antediluvian King Saurid had the following message inscribed:

> I, Saurid, the king, have built the pyramids at this time and have completed their construction in six years. Those who come after me and believe that they are kings like me, should try to destroy it in six hundred years: and it is known that it is easier to destroy than to build. When they were finished, I had them covered with brocade, may they cover them with mats. . . . When King Saurid ben Sahluk died, he was buried in the eastern pyramid; Hugib, however, in the western one; and Karuras in the pyramid that is made on the bottom with stones from Aswan and at the top with Kaddan stones.

These pyramids have gates below the ground which lead into a vaulted corridor. Each corridor measures one hundred and fifty yards. The gate to the eastern pyramid is on the north side, the gate of the western pyramid is on the west side, and the gate to the vaulted corridor of the stone-covered pyramid is on the south side. The pyramids contain unimaginable treasures of gold and emeralds.

The man who translated this text from the Coptic language into Arabic added the dates up to the sunrise of the first day of Toth—it was a Sunday—in the year 225 of the Arabian calendar and came up with 4,321 solar years. When he examined how much time had passed after the Flood up to that particular day, the result was: 1,741 years, 59 days, $13\frac{4}{5}$ hours, and $\frac{59}{400}$ hours. He subtracted this from the previous sum and was left with 399 years, 205 days, 10 hours, and $\frac{21}{400}$ hours. He then realized that this text was written so many years, days, hours, and partial hours *before* the Flood.

The *Khitat* lists several Arabian sources, one after another, which contradict each other as to the construction date of the pyramid. The following is only one example:

> Abu Said al-Balhi states: On the pyramids was an inscription written in their language. It was understood, and it read: "These two pyramids were built when the 'Falling Vulture' was in the tropic of cancer." So they calculated from this point to the Hegira of the prophet and came up with two times 36000 solar years.

Who was this provident King Saurid? Is he a mythical fantasy figure, invented in a dream world of wishes and desires, or can he be realistically placed? The *Khitat* states that he "was Hermes, whom the Arabs call Idris." It is said that he was personally instructed by God in the science of the stars and that God told him that a catastrophe would befall the earth, but that a part of it would survive and would have a need for scientific knowledge. Consequently, Hermes (a.k.a. Idris, a.k.a. Saurid) had the pyramids built. The *Khitat* is even more specific:

> There are people who say: the first Hermes, who was called the "triune" because of his function as prophet, king, and wise man (he is the one whom the Hebrews call Enoch, the son of Jared, the son of Mahal'aleel, the son of Ca-i'nan,

> the son of Enos, the son of Seth, the son of Adam—blessed be his soul—and that is Idris), read in the stars that the Flood was about to come. So he had the pyramids built and had hidden inside of them treasures, learned writings, and all those things which he feared might get lost or disappear, so that they would be protected and well preserved.

People in the West are not used to thinking in dimensions preceding the Great Flood and so we ask, in our confusion, why the Arabian historians insist on dating the pyramids to a time before the Flood. Muhammad ben Abdallah ben Abd al-Hakam submits an excellent argument. "In my opinion, the pyramids can only have been built before the Flood; if they had been built after, people would know more about them." A very good argument, indeed. Irrefutable.

The statement in the *Khitat* that Enoch, in the Old Testament, is identical with Hermes and Idris is very exciting. Much can be derived from it. Enoch (a.k.a. Hermes, a.k.a. Idris, a.k.a. the *Khitat*) is not the only source to identify Saurid as the builder of the pyramids. Ibn-Battuta, a fourteenth-century Arabian explorer and writer, also assures us that Enoch had the pyramids built before the Flood "to preserve in them books about science and knowledge and other valuable objects."[11]

My Friend Enoch

Who is this Enoch? My faithful readers know him from my earlier work.[12] Thus, I will keep my description short.

In Hebrew, the name "Enoch" means "the consecrated one, the judicious one, the knowing one." Moses calls him the seventh of the ten forefathers, an antediluvian patriarch, who has been overshadowed by his son Methuselah for thousands of years. The book of Genesis tells us that Methuselah lived to be 969 years old—hence the expression "as old as Methuselah." The Old Testament mentions Enoch only briefly, although he does not deserve to be shortchanged in such a way. Enoch is the author of several exciting first-person narratives. The books of Enoch are not part of the Old Testament. The church leaders did not understand Enoch and therefore precluded him from public viewing. Fortunately, the Ethiopian Church did not follow these orders. Enoch's writings were included in the Old Testament canon of the Abyssinian Church and have been in its register of holy scriptures ever since.

At present, two versions of Enoch's books exist, which are essentially similar: the Ethiopian and the Slavic Enoch books. Highly academic text comparisons proved that they are based on the same original, written by a single author. Those who attempt to interpret Enoch's writings from a stubbornly theological viewpoint encounter a maze of curious statements. However, if we strip the text of its ornamental embellishments and flowery symbolism, we are left with an incredibly dramatic skeleton narrative, without changing its meaning one bit.

The first five chapters of the book of Enoch announce a world judgment. Chapters 17 through 36 describe Enoch's travels to different worlds and distant constellations. Chapter 37 through 71 contain various kinds of parables which the "heavenly beings" related to the prophet. Chapters 72 through 82 report in exact detail information about solar and lunar orbits, leap days, stars, and universal occurrences. The remaining chapters record conversations between Enoch and his son Methuselah, whom Enoch warns of the coming Flood. In the happy ending, Enoch rides into the sky on a fiery chariot.[13]

The Slavic book of Enoch[14] features additional passages which are missing from the Abyssinian version. The Slavic book reports how Enoch was contacted by the heavenly beings.

> The books of the holy parables of Enoch, the wise man and great writer, whom the Lord took in and loved, so that he may see the dwellings of the Highest. . . . In the first month of the 365th year, on the first day of the first month, I, Enoch, was in my house alone . . . and two very tall men appeared before me, as I had never seen on earth. And their faces were shining like the sun, their eyes burning like torches, and fire was spewing from their mouths; their feathers of different looks, their feet purple, their wings more radiant than God, their arms whiter than snow. And they stood at the head of my bed and called me by my name. I awoke from my sleep and clearly saw those men standing next to me. And those men spoke to me: "Be brave, Enoch, . . . you are going up to heaven with us today. And tell your sons and all the children of your house everything, how much they shall do in your house on earth without you, and no one shall look for you until the Lord returns you to them. . . ."

Enoch is lifted from the earth. There he is introduced to a number of "angels."

He is handed a device for "speed writing" and told to write down all that the angels dictate to him. "O, Enoch, see the writing on the heavenly tablets, read what they say, and remember it all in detail."

Three hundred and sixty books are created in this way. They are the gods' testament to mankind. After many weeks, the strangers return Enoch to his house, but only so that he can bid farewell to his loved ones for good. He bequeaths the books which he wrote to his son Methuselah and gives him unmistakable orders to keep the books safe and to leave them for future generations of this world. What has happened to the books? All of them have vanished, except for the Enoch books mentioned above, and are presumed lost.

Whenever the topic of Enoch comes up in discussions and I suggest that the antediluvian prophet had the privilege to receive instruction inside an extraterrestrial spaceship, I am told that he would have needed some kind of space suit. Would he really have? Astronauts move freely without space suits inside our shuttles and space stations. Only the danger of unwanted contamination with viruses and bacteria would have forced the extraterrestrials—and Enoch—to protect themselves. What does Enoch, the attentive student, report?

> And the Lord said to Michael: Approach and disrobe Enoch of his earthly garments and anoint him with a good ointment and clothe him in the robes of my splendor. And Michael did as the Lord had told him: He anointed me and clothed me. And the appearance of the ointment was more than a great light and its greasiness like good dew and its aroma like myrrh and shining like the rays of the sun. And I looked at myself, and I was like one of his splendid ones, and there was no difference in appearance.

A curious description indeed. Should the one true and universal god supposedly give orders to have Enoch anointed with a particularly greasy, intensely aromatic ointment? Human beings have always had a very peculiar smell.

Are there connections between the Old Testament prophet Enoch and the unknown King Saurid, whom the Arabs credit with the building of the pyramids.

a) Both lived prior to the Great Flood.
b) Both were warned of the coming Flood by gods.
c) Both authored books about all the sciences.
d) "God personally" instructed both of them in astronomy.
e) Both gave orders that their works be preserved for future generations.

In contrast to these parallels, there are also some serious discrepancies. Saurid is said to be buried in a pyramid; Enoch left the earth in a heavenly vehicle. And no indication is found in the existing books of Enoch that the biblical patriarch commissioned the construction of pyramids.

Undoubtedly, we can also find parallels between Enoch, Saurid, and Hermes, the Greek messenger of the gods. But Hermes is not antediluvian, and neither did he build any pyramids.

Professional experience has taught me to see more behind traditional folk tales than just human imagination and creative storytelling. Myths can be screened in such a way that the ornamental additions are diluted and the fundamental truths enhanced. Around 700 B.C., Hesiod, a Greek poet, wrote in his "Myth of the Five Human Races," that in the beginning the immortal gods, "Cronus and his colleagues," created man: "the exalted race of those heroes, called demigods, who lived on the infinite earth in a time before us. . . ."[15]

Demigods are also demihumans, earthly beings with extraterrestrial genes. Hermes, Enoch, Idris, and Saurid—all of them were members of this elite group. For all of them, the designation "a very long time ago" is appropriate. Sources link all of them with "written books" that were hidden. This common denominator applies to Saurid, Idris, and Enoch, as well as to many other teachers of the human race, including the demigods described by Hesiod.

If myths were really as nebulous as they are often made out to be, they would not yield any useful information. It has always been more convenient to accept someone else's teachings—proven or not—than to use one's own head and to take the time to find similarities in mythical contents. I am not talking about an academic comparison of myths here; that would require a much more extensive approach. I am still pondering the construction of the Great Pyramid and the possibility that ancient documents may lie inside the pyramid which could revolutionize our entire religious thinking, as well as our notions about early human history and evolution.

My friends the Egyptologists see no reason to discount Pharaoh Khufu as the builder of the Great Pyramid. The chronology of dynasties leaves no room for an additional ruler and possible builder after Khufu. They all erected their own, precisely dated sacred monuments. Moreover, the names of the Egyptian kings are recorded on the "Torino Papyrus," a document dated to the thirteenth century B.C. which today is stored in Torino. Egyptologists also found the names of kings listed in the temple of Seti I in Abydos, as well as on several walls of the temple complex of Karnak. I do not begrudge Egyptologists their scrupulously documented, detailed lists. They have nailed down all the Egyptian rulers.

Documented Millennia

How does it look *prior* to Khufu? The recorded dynasties begin around 2920 B.C. with a so-called Thynite king named Menes. (Min and Hor Aha are also mentioned.) Under Menes, however, the Egyptian state must have been rather well organized already, because Menes led military excursions past the borders of his empire. He also ordered the course of the Nile to be altered south of Memphis. Such tasks cannot be accomplished out of the blue. Menes must have had predecessors.

The problem with dates is this: Christians count the years since the birth of Jesus Christ; the Romans counted the years after the founding of the city of Rome in 753 B.C. With ancient Egypt, however, we have no fixed calendar starting point that can be translated into numbers. We are floating on pudding; there is nothing solid we can sink our teeth into. The experts went through great pains to reconstruct their chronology after Menes from datable findings, buildings, and astronomic calculations. Their dates are rather accurate, with a few exceptions, but they tell us nothing about the time prior to the first dynasty.

This is where legend enters the picture. To the amazement of scholars, the legends also list precisely the names and reigns of various kings and support them with numbers, but archaeology lacks the corresponding monuments or artifacts. How should we interpret names and dates that reach back thousands of years into the past, but are not documented with physical evidence? They are relegated to the realm of myths.

It is believed that the Egyptian priest Manetho wrote eight books, including one book about the history of Egypt. They list the names and dates of prehistoric kings, even back to the times of gods and demigods. Where did Manetho, who lived in the third century B.C., get his information? It was customary, even back then, to record time on the basis of extraordinary events. What ensued were lists of dates, which grew into annals. The priests guarded and copied these annals so that they could recite the heroic deeds of certain individuals and the superb admirable accomplishments of the gods. Even later on, in the heyday of the pharaonic empire, when the annals no longer recorded exact calendar dates, it was customary to consult the annals if extraordinary occurrences took place. The Egyptians wanted to check if similar events had happened at some time in the past. Allegedly, Ramses IV found his name written in golden letters on a tree when he visited Heliopolis. Immediately, the annals were consulted and checked all the way back to the beginning of the pharaonic reign, as far back as history was recorded, but no mention was made anywhere of a similar event.[16] Similarly, the annals were searched for evidence of unusual climatic disasters or the expected return of the gods.

Manetho, the priest, was able to study such annals in researching his books. He writes that Hephaestus was the first ruler of Egypt and that he invented (or brought?) fire. He was followed by Cronus; Osiris; Typhon, one of Osiris' brothers; and then Horus, the son of Osiris and

Isis. Eusebius states that, after the gods, their offspring ruled the earth for 1,255 years, that then other kings reigned for 1,817 years, followed by thirty kings from Memphis, who dominated for 1,790 years, and finally by ten Thynite kings, who ruled for 350 years. The rule of the spirits and divine offspring spanned 5,813 years.[17]

Eusebius, the bishop of Caesarea, who adopted these numbers from Manetho,[18] emphasizes that these are lunar years, but they would still date back more than thirty thousand solar years B.C. Not surprisingly, Manetho's numbers are challenged by scholars, because there is no fixed reference point on which to base the dating process.[19]

When time frames encompass tens of thousands of years, archaeologists get uncomfortable. Manetho's figures are cut down to lunar years; he is accused of exaggerating, because it would have been in his interest, as a priest, to enhance the credibility of the priesthood by giving it an ancient history. Those critics who do not doubt Manetho's integrity suggest that he copied old annals which were already full of exaggerations. It is difficult to understand, however, why other ancient writers, who were neither priests nor Egyptians and who had nothing to gain by exaggerating the numbers, quote the same "impossible" dates.

Diodorus Siculus, the historian who created a forty-volume historical chronicle and who is often skeptical, reports in his first book that the ancient gods founded many cities in Egypt.[20] He also claims that some of the gods' offspring became kings of Egypt. In those ancient times, the predecessor of Homo sapiens was a primitive creature. Diodorus asserts that only the gods put an end to man's cannibalistic habits. The gods, according to Diodorus, taught man about the arts, mining, the manufacture of tools, agriculture, and the making of wine.

The helpful divine beings also gave language and writing to mankind.

> For this god [Hermes, a.k.a. Enoch] was the first to bring language to perfection; he named many nameless things, invented the alphabet, and ordained ceremonies governing divine worship and sacrifices to the gods. He was the first to perceive order in the stars and to discern the nature and harmony of musical sounds. . . . In short, they hold him to have been the sacred scribe of Osiris.

It is plain as day. Unbeknownst to Diodorus, Enoch, too, is described as a "sacred scribe." Just as Diodorus, who has never heard of the biblical patriarch Enoch, relates in his first-person narrative that the "guardians of heaven" taught many things, good and bad, to the people on earth.[21]

> The name of the first is Jequn; he is the one who seduced all the children of the angels, brought them down to earth and seduced them with the daughters of men. The second is named Asbeel; he gave bad advice to the children of the angels that they contaminated their bodies with the daughters of men. The third is named Gadreel; he is the one who taught the humans all kinds of deadly moves. He also seduced Eve and showed the mortals murderous instruments, armor, shields, battle swords, and generally all kinds of murderous weapons. . . . The fourth is Penemue; he taught the humans the difference between bitter and sweet and showed them all the secrets of their wisdom. He taught man how to write on paper with ink.

Why do we discount such stories, which, thousands of years ago, were an integral part of historical knowledge? Do our historical interpretations, as far as they delve into the times before Pharaoh Menes, offer more reasonable facts? Are there any convincing arguments against Diodorus? I have been told that I am being too simplistic, that we cannot rely on Diodorus alone. True. But that is the curse of our specialized sciences. An Egyptologist knows nothing about ancient Indian sources; a Sanskrit scholar is unaware of Enoch or Ezra; an Americanist is as ignorant about the Rig-Veda as a Sumerologist is about the Mayan god Kukulkán, etc. And if some smart scientist actually does engage in comparative studies, it is always from a narrow-minded theological or psychological point of view. The evidence submitted by Diodorus was confirmed thousands of years ago, internationally, even if the various authors used different names and details in their stories. In essence, all ancient chroniclers from all parts of the world present the same story. I wonder why we refuse to believe these people. I know that truth never triumphs, but its opponents gradually become extinct. Diodorus stated very matter-of-factly that the Egyptian god Osiris also founded a number of cities in India. The truth of this statement is so obvious to me that I have no patience for any academic discussion about it. What are the dates reported by Diodorus?

> They record that ten thousand years elapsed from the time of Isis and Osiris until the reign of Alexander, who founded the city in Egypt named after himself; and some write that it was just short of twenty-three thousand years.

A few pages later, Diodorus describes the battle between the Olympian gods and the giants. Diodorus maintains that the Greeks must be mistaken in placing the birth of Heracles a mere generation before the Trojan War, asserting that "it was rather, they [the Egyptians] argue, nearer the first origin of mankind: for from that the Egyptians reckon more than ten thousand years, but from the Trojan War less than twelve hundred."

Diodorus knows what he is talking about, because later he relates the Egyptian dates to his own visit of Egypt. He writes that "the first rulers in Egypt, for nearly eighteen thousand years, were gods and heroes, and that the last of the gods to hold sway was Horus, son of Isis; but they say the country has been ruled by mortals now for close to five thousand years, down to the One Hundred and Eightieth Olympiad, at the time of my visit to Egypt."

Diodorus did his homework. He studied the ancient sources that were available and talked to knowledgeable people. We did not. We destroyed the old libraries in the name of whatever religion was prevalent at the time; we burned precious old documents and murdered the learned and wise men of their cultures. The library of Carthage contained five hundred thousand documents. Where are they? Burned! The *Sibylline Books* and the *Avesta of the Parsees*, written in golden letters. Where are they? Burned! The libraries of Pergamum, Jerusalem, Alexandria, with millions of documents. Where are they? Burned! The priceless manuscripts of the Central American cultures. Where are they? Burned! Our pyromaniacal past shines much more brightly than the dim wits of some of our "revolutionary" thinkers.

Herodotus and the 341 Statues

Herodotus, who visited Egypt several centuries before Diodorus, also illustrates the lengthy time frame of Egyptian history in the second book of his *History*. He recalls how the priests in Thebes personally showed him 341 statues, each of which was representative of one generation of high priests for the past 11,340 years.[22]

> . . . the custom being for every high-priest during his lifetime to set up his statue in the temple. As they [the priests] showed me the figures and reckoned them up, they assured me that each was the son of the one preceding him; and this they repeated throughout the whole line, beginning with the representation of the priest last deceased, and continuing till they had completed the series.

Why would the priests have lied so shamelessly to Herodotus about the 11,340 counted years? Why do they emphasize that no gods had been among them for 341 generations? Why do they use statues to illustrate their precise times and dates? Herodotus, who was far from being gullible, assures us that the priests proved to him that they were telling the truth. He carefully separates fact from fiction.

> Thus far I have spoken of Egypt from my own observation, relating what I myself saw, the ideas that I formed, and the results of my own researches. What follows rests on the accounts given me by the Egyptians, which I shall now repeat, adding thereto some particulars which fell under my own notice.

According to our academic doctrines, Menes was the first pharaoh of the first dynasty (around 2920 B.C.). Herodotus' report that Menes changed the course of the Nile above Memphis is widely accepted, while another fact, related by Herodotus only eighteen lines later, is just as widely ignored: "Next, they read me from a papyrus the names of three hundred and thirty monarchs, who (they say) were his [Menes'] successors upon the throne."

Is there really no room for a monarch who could have built the pyramids among the three hundred and thirty-five kings succeeding Menes? Moreover: In light of the statues which the priests displayed before Herodotus, each of which represented one generation of high priests, the question of lunar years becomes academic. "You may deceive all the people part of the time, and part of the people all the time, but not all the people all the time" (Abraham Lincoln).

The Eyes of the Sphinx

Once upon a time, there was an Egyptian prince who liked to hunt near Memphis, where the great pyramids are. One day, he sat down, exhausted, in the shade of the sphinx and fell asleep. Suddenly, the "great god" opened his mouth and spoke to the sleeping prince, as a father speaks to his son:[23]

> Look at me, my son Thutmose. I am your father, the god Re-Herakhty-Atum. I will give to you the kingdom . . . the treasures of Egypt and the great tributes of all nations shall be yours. For many years, my eyes have been upon you, and my heart. The sand of the desert on which I stand is encroaching upon me. Promise me that you will fulfill my wish.

The prince became Pharaoh Thutmose IV (1401–1391 B.C.). In his first year as king, he addressed his divine father's plea. He had the sphinx dug out from under the sand. Thutmose inscribed the touching story of his dream on a stele, which is located between the sphinx's front paws.

The sphinx—is it male or female? No one knows for sure, because the experts cannot agree whether the colossal statue originally bore male or female traits. Maybe it had both. In any case, Thutmose did not save the sphinx for long. Again it was swallowed by the sand. Then the Ptolemies dug it up again, and the sand sucked it down again.

Historically well documented is the excavation of 1818 at the hands of Giovanni Battista Caviglio, the same man who was chased off by Howard Vyse. Caviglio discovered an anteroom with a stone floor between the lion paws, which was divided up by a passageway in which a stone lion was found. Seventy years later, the sphinx had to be dug up again, this time by Gaston Maspero, then director of the Egyptian Department of Antiquities. Forty years later, the sphinx disappeared under the sand once more. In the time of Herodotus, this strange and mysterious statue must also have been invisible, because the Father of History makes no mention of it.

What is the sphinx? A lion's torso, fifty-seven meters long and twenty meters high, made from one gigantic boulder, with an enigmatic head and a veil over the back of its head. Kurt Lange, an Egyptologist, calls the statue "the monumental symbol of royal power."[24] What is it supposed to represent? What does it symbolize? What is its purpose? What was it intended for? There is no answer to these questions. The centuries have eaten away at the monument; possible inscriptions and a statue which the sphinx was cradling at one time have disappeared.

Richard Lepsius wondered about the meaning of the sphinx, which during his visit was half buried in sand. "Which king does it represent?" Lepsius asked,[25] while wondering too, "If the statue depicts King Khafre, why does it not carry his name?"

The eyes of the sphinx are wide open. Expectantly, yet calmly, it surveys the tiny people below deliberately, confidently . . . and even a little mockingly, it seems to me, head and shoulders above the world. The experts agree on one thing: The sphinx of Giza is the oldest sphinx, the mother, the model on which all later imitations were based. It is attributed to Pharaoh Khafre not because of any solid evidence pointing in that direction, but merely because the name "Khafre" was found on a crumbled piece of the Thutmose stele—provided one *wants* to see the name "Khafre" on it. Thutmose lived more than one thousand years after Khafre. He is the only one who could tell us in what context the name "Khafre" appeared on his inscription.

Gaius Plinius Secundus writes:[26]

> Over against the said pyramids there is a monstrous rock called Sphinx, much more admirable than the pyramids. . . . within it the opinion groweth, that the body of king Amasis was entombed; and they would bear us in hand, that the rock was brought thither, all and whole as it is.

Plinius also claims that, despite its marvelous stature, the sphinx was largely ignored by ancient writers and that it was from a single natural stone. The red face of the monster was worshiped like a god.

Egyptologists do not list a king named Amasis in their records, and no tomb has been found under the sphinx so far. Maybe Plinius' Amasis is identical to an Amasis mentioned by Herodotus. That would bring us right back to the realm of myth because Herodotus writes that, according to the Egyptians, seventeen thousand years had elapsed until the reign of Amasis.

The sphinx and the pyramids have been inextricably related throughout history. Both are monumental in size—and nameless. No one would have been able to chisel a hybrid monster, fifty-seven meters long and twenty meters high, out of the rock in the wink of an eye. The fabulous statue could not have been created without detailed plans and blueprints or without scaffolding. On the pyramids, or inside, we would expect inscriptions such as "I, Pharaoh XY, created this building." On the sphinx, an appropriate inscription would be something like "I, goddess/god XY, am guarding this burial site." Or, "For eternity, I want to remind mankind of . . ." Why did neither the pyramids nor the sphinx receive a label? Did some bombshell surround these buildings that, even back then, was deliberately obscured? Is the fact that these monuments are nameless more than just carelessness or malice on the part of later generations? Was it intended that way? Diodorus Siculus drops a bomb on us when he claims that some of the original gods were buried on earth. I beg your pardon? And where? if I may ask.[27]

Diodorus reports that the information which was circulated regarding the burial of these gods was contradictory because the priests were prohibited from spreading their knowledge of these matters; they did not want the people to know the truth. Anyone who promulgated the secret knowledge about these gods among the masses faced great danger.

Brief as it is, this passage has explosive implications. Gods are buried somewhere on this earth! The high priests were aware of this but were not allowed to spread the news. Couldn't it be that one of the god-kings was laid to rest under the Great Pyramid? Wouldn't it be irrelevant in that case if his name was Saurid, Idris, Hermes, Enoch, or anything else?

If . . . if the Great Pyramid was built by a god-king or the offspring of a god . . . if this happened before Khufu's time . . . if the Great Pyramid holds secret books and important devices . . . if one of these god-kings is indeed buried inside the pyramid, then the namelessness is intentional. Diodorus solved the mystery: Disseminating the news about the tombs of the gods was strictly forbidden.

And the sphinx? In this scenario, the sphinx becomes a grand memorial to the relationship between the terrestrial and the extraterrestrial element, between the earthly creature and the divine intellect. It is the petrified symbol of the union between flesh and analytical thinking, between raw, primitive strength and sublime culture. For thousands of years, the sphinx has been smiling, sarcastically and sensitively. The eyes of the sphinx are watching our progress, benevolently, until the day when our own eyes are opened. This day cannot be far off. The hidden chambers and tunnels inside the Great Pyramid have been located.

A Pharaoh Disappears

There is yet another highly captivating mystery in store for us, this one set up by a pharaoh who, irrefutably, lived sixty years before Khufu: the third-dynasty ruler Sekhemkhet (2611–2603 B.C.). This monarch had a pyramid of his own constructed southwest of the step pyramid of Saqqâra, but apparently the pyramid was never completed. As a matter of fact, it only climbed to a height of about eight meters.

Over thousands of years, this pyramid disappeared completely below the desert sand. It was not until 1951 that it was resurrected, by the Egyptian archaeologist Zakaria Goneim.

Dr. Zakaria Goneim was considered a highly intelligent and talented archaeologist, the very opposite of a reclusive or stubborn scholar. His seminars and excavations were saturated with friendly humor; he always had a sensitive and open mind for the questions submitted by his students. It was one of his great talents to enliven the bare bones and ruins he excavated with interesting stories. When Zakaria Goneim discovered the entrance to a corridor under the pyramid of Sekhemkhet, he hoped and prayed that the sepulcher had survived the centuries intact.

It took much sweat and many years to dig through the sand and rock formations. Zakaria Goneim located another corridor filled with thousands of animal bones, some of them from gazelles and sheep. He also found sixty-two broken tablets with text fragments from the year 600 B.C. Someone had deposited them under the pyramid two thousand years after the death of Pharaoh

Sekhemkhet. Finally, in the last days of February 1954, the excavating team reached the entrance to the actual tomb, buried far beneath the desert. Zakaria Goneim generously stepped aside to let the minister of culture open the sepulcher officially. It was March 9, 1954 when the final, decisive blow struck the entrance gate.

Through one last tunnel, the men crawled into an underground hall, hewn roughly out of the rock, just like the "unfinished burial vault" under the Great Pyramid. In the center of the room stood a marvelous polished sarcophagus made from white alabaster, a kind of marble. The remains of a flower bouquet were scattered on the north end of the sarcophagus, a last farewell to the deceased pharaoh. Zakaria Goneim, realizing at once the significance of this find, immediately ordered his assistants to cover the plant dust carefully. The substantial amount of plant dust was proof that the sarcophagus had not been touched. The archaeologists and workers cheered, danced, and jumped with joy in the subterranean hall. Finally, an untouched sarcophagus!

In the days that followed, the splendid sarcophagus was examined carefully. There was not a smidgen of evidence that it had been forced open within the past forty-five hundred years; there was no indication that anyone had even *attempted* to open it. Undoubtedly, Pharaoh Sekhemkhet was laid to rest inside the sarcophagus. The crumbled flowers cemented the archaeologists' faith. The magnificent sarcophagus—which looked as if it was made out of one piece of alabaster—was not only extraordinary because of its material and its creamy white color, but also because it featured a sliding door, which allowed it to be sealed off hermetically. Ordinarily, sarcophaguses are covered with a lid, which is placed on top of the sarcophagus casket. Not in this case. The sarcophagus of Sekhemkhet was equipped with a sliding door, much as an animal cage is, which could be raised in the front; smooth tracks had been miraculously carved out of the alabaster. It was a unique and incomparable work of art, the most beautiful and also the oldest sarcophagus Egyptologists had ever laid eyes upon.

Zakaria Goneim hired a special Sudanese police force to guard the vault day and night. Sudanese policemen were known for their stubborn adherence to orders. No one was allowed to enter. Nothing inside the tomb was to be touched until the official opening of the sarcophagus.

The day finally came. It was July 26, 1954. Representatives of the Egyptian government, a select group of archaeologists, and dozens of journalists from all over the world had been invited. Film and photo cameras had been set up. Spotlights were trained on the sarcophagus. Chemicals were on hand in case something needed to be preserved right away. Overcome with intense feelings of hope and happiness, Zakaria Goneim took a last good look at the sarcophagus. Then he gave the order to open the alabaster coffin.

Two workers wedged their knives and chisels into the barely visible cracks on the bottom of the sliding door. Ropes were tied to the door, and some other workers, positioned on top of the sarcophagus, pulled with every muscle they could muster. For two hours, everyone labored, trying to hoist the sliding door. Finally, it opened just a crack. The alabaster was moaning and groaning as the door was lifted a few centimeters. Wooden wedges were placed under the door. Quietly, in breathless anticipation, the journalists and archaeologists watched as the opening grew one centimeter at a time.

Zakaria Goneim was the first to kneel in front of the sarcophagus, shining his lamp into the coffin. Perplexed, incredulous, he pointed his light through the opening—and found the sarcophagus empty!

The archaeologists' world was turned topsy-turvy. The journalists, feeling cheated out of a story, left the burial site gravely disappointed. For days, Zakaria Goneim kept returning to the sarcophagus, but it remained empty. The magnificent alabaster coffin was squeaky clean.

The Sleeping Dead?

Now what? Had the mummy of Sekhemkhet just left without a trace, or was the pharaoh never buried at all? The latter option, although theoretically conceivable, does not conform to the facts at all.

Remember this: The sarcophagus was completely sealed; it had not been touched in thousands of years. On top of the sarcophagus was a last flowery farewell, possibly from his lover, who'd had the privilege of accompanying her master down to the vault.

As I was standing in the subterranean hall, taking pictures of the incomparable sarcophagus and the flower dust from all possible angles, some impromptu thoughts flashed through my mind. Although they might be properly put in the science-fiction category, they cannot be totally dismissed. I was not willing to accept the empty sarcophagus without questioning and to bury my sentiments in foggy ignorance.

What had Diodorus Siculus reported two thousand years earlier? That some of the original gods had been buried on earth? I was standing inside the vault carved out of the rock ages ago, prior to Khufu. A harangue of contradictions echoed from the stone walls, as if Hermes himself were sitting there, laughing at my confusion. On the one hand, a magnificent sarcophagus, unequaled in its beauty; on the other, a roughly hewn vault in the rock, without polished ceilings or monoliths. The powerful yet delicate appearance of the sarcophagus was out of place inside this primitive hole in the rock.

It was a situation similar to the "unfinished burial vault" under the Great Pyramid. Was I standing in front of the sarcophagus of a legendary ancient king? Was it the resting place of a divine offspring? Evidently, it was not a permanent resting place, because Zakaria Goneim did not find a corpse. Maybe it was to be a mere stopover, for a few decades or a handful of centuries at most, until his space-traveling companions could pick him up and resurrect him? Absurd? Our scientists are studying the possibility of putting future astronauts into a deep sleep for long journeys. The idea is not that unrealistic. Had XY, the son of the gods, consumed all his earthly time? Could it be that he was seriously ill? Had he accomplished his mission on earth? Was it merely a matter of inducing a deep sleep with the help of the right drugs, then waiting for the mother ship to return so it could locate him and take him aboard? Was that the reason the sepulcher did not need to be, indeed could not be, perfected with monoliths? Clearly, the humans, in their reverent and dedicated zeal, would not have stopped polishing the monoliths until they fit perfectly. It would have taken years of labor inside the "sleeping chamber," and that was out of the question. Once deep sleep set in, no one—no mason, no priest—would be allowed to enter the underground chamber. By royal decree, the vault containing the sarcophagus had to remain anonymous and be forgotten. That is why the priests, according to Diodorus, were prohibited from disseminating their knowledge about these things.

The Origins of the Rebirth Concept

Did this predominant idea of a rebirth arise when the original gods were preparing for their deep sleeps? Did later pharaohs merely copy what the priests, with their secret knowledge, had known for years and subsequently told their top bosses, the pharaohs: that dead bodies are asleep, that they will be picked up by the gods and carried off into space? Is this the real reason behind the later pharaohs' belief that they had to keep material treasures like gold and precious stones handy in their tombs, so that they would be able to pay the resurrection crew? Is this the reason why the pyramid texts describe in such flowery and optimistic terms the deceased pharaoh's future journey to the stars?

These questions, I grant you, are speculative, but they are legitimately provoked by the facts disclosed by ancient sources. After all, our present knowledge would not exist without the past.

Even though no one has yet found an "ancient sleeping king" or the mummy of a divine offspring, there is evidence that points to their previous existence. Humans are great imitators. They always follow—to the present day—some kind of model. Not true, you say? Then what else but imitation of a pretty model are the annual fashion trends followed by thousands? Humans have copied scepters and crowns; technical devices, as the cargo cults have shown; and ideals of beauty. It would be truly surprising if they did not also imitate the appearance of their gods.

What example of irrational behavior can we find that has had international appeal, allowing us to draw worldwide parallels?

The deformation of skulls! They are the most horrendous example of human vanity and are so typical of human nature. Without electronic communication devices, jet planes, or TV satellites, our prehistoric ancestors engaged in a cult of skull deformation all over the globe. The deformations begin at the temples and arch back from the forehead, like the torso of a wasp. Often, the back portion of the head is three times as large as a regular skull.

We know that the Inca priests of Peru selected boys at a very young age and squeezed their small, undeveloped heads between padded boards. Ropes were threaded through hinges and the space between the boards gradually reduced. Some of the children must have survived the torturous, indescribably painful procedure, or else we would not have the deformed skulls of adult males as evidence.

What perversion motivated our ancestors to flatten the tender heads of their own children to such an extent? Naturally, archaeologists whom I consulted in this matter were unable to offer a sensible reason. They claimed that there must have been some useful purpose; maybe it was easier to wear headbands with a deformed skull. However, a normal head with a normal forehead is more suited for carrying heavy loads with a headband than is an elongated skull. The archaeologists also suggested that the deformed skull corresponded to some kind of aesthetic ideal or that it distinguished a certain social group from others.

But, friends: Skull deformations are not a Peruvian specialty! They are found in North America, Mexico, Ecuador, Bolivia, Peru, Patagonia, Oceania, in the Eurasian tundra, in Central and West Africa, in the Atlas Mountains, in early Central Europe, and, of course, in Egypt.[28]

Proof

Why? The children had to be deformed so that their skulls would resemble the skulls of the ancient gods. All around the world, humans had encountered the intelligent and majestic beings. Presumptuous copycats everywhere wanted to look like the divine beings. Shortly, the priests started the barbaric tradition of elongating their heads to resemble the gods. It was a great way to impress their fellow men and women. "Look at him, he looks like, moves like a god. He must have special knowledge and, consequently, special power over ordinary humans." If skull deformation was limited to only a single culture, one might be able to find local origins for its existence. That is not the case. In pictures worldwide, the elongated skull is an attribute of the gods. The gods of Egypt and their offspring beckoning from statues and temple walls with their oversized heads are irrefutable proof.

I did not make up the ancient gods, the teachers who came from space, and I am not the father of divine offspring or god-kings. The confusing data about those obscure prehistoric times are not a figment of my imagination, and neither is the claim that scientific books and precious objects are hidden inside the pyramids. I am not responsible for the fact that the pyramids and sphinxes are not labeled, and it is not my fault if a phenomenal sealed, yet empty, sarcophagus surfaces inside a subterranean vault. I have assembled this conglomeration of stories and facts for discussion purposes, because scholarly efforts follow a one-track course and I would like to inject some fresh blood into those academic veins.

When I view all the evidence we have collected about these long-gone days, I am reminded of a quote by Michel Eyquem de Montaigne, with which he culminated a speech in front of a gathering of eminent philosophers: "Messieurs, I simply picked a bunch of flowers and added nothing but the thread that binds them."

✧ 5 ✧

The Great Pyramid: The Latest and Greatest

"The more one knows, the more one is likely to doubt."

VOLTAIRE (1694–1778)

It has been quite apparent on a number of occasions that some scientists have a low regard for the intelligence of the general public and that public opinion is often shamelessly manipulated by the media. This has been amply evidenced in the past two years by the coverage surrounding Khufu's Great Pyramid in Egypt. On March 22, 1993, at precisely 11:05 A.M., a sensational event took place at the pyramid. The unexpected, the unthinkable, occurred, an event so unfathomable and incomprehensible to classically trained archaeologists that it shattered and totally devastated the established framework of Egyptology. But the shock waves were suppressed, contained, trivialized, and an even greater sensation—one of the most important events of human history, comparable to the discovery of an extraterrestrial intelligence—was swept under the rug.

What happened?

Rudolf Gantenbrink, a German engineer born in Menden on December 24, 1950, delivered a stroke of genius. Sixty meters into a previously unknown tunnel inside the pyramid, a small robot with state-of-the-art equipment discovered a door with two metal braces. The robot had been trudging through the narrow shaft for two whole weeks, frustrated again and again by various obstacles. Repeatedly, the robot had to be ordered back to its starting point with the help of specially designed electrical wires. Minute adjustments were performed on the technological wonder before it was sent back into the age-old tunnel.

Gantenbrink's caterpillar-type robot weighs six kilograms and measures thirty-seven centimeters in length. The miraculous gadget is powered by seven independent electrical motors with remote-controlled microprocessors. The front features two small halogen lights and a Sony miniature video camera that can be moved both horizontally and vertically. Despite its lightweight aluminum construction, the robot has a maximum tow capacity of forty kilograms, thanks to its specially designed rubber tracks, which can grip both the floor and the ceiling.

All the innovative ideas behind the development of this unique vehicle come from Rudolf Gantenbrink. He personally built the robot to exacting mechanical standards over a period of several months. He invested an immeasurable amount of time, labor, and money—more than $300,000—in the construction of his technological masterpiece. Technical support was provided by the Swiss company ESCAP in Geneva (specialty motors), by

HILTI AG in Vaduz, Liechtenstein (drilling technology), and by the GORE company in Munich (special wiring). Gantenbrink's robot is an inspiration to all those perennial doubters who feel that their goals can never be accomplished. It can't be done? It *can* be done, with a combination of intelligence, technology, and determination.

What motivated Rudolf Gantenbrink to venture into the Great Pyramid? Surely, everyone knew that there was nothing more to discover there. Torsten Sasse, a radio and TV journalist from Berlin, interviewed Rudolf Gantenbrink.[1]

> The whole thing started when I was staying in Egypt during the Gulf War. I suggested to Professor Stadelmann [of the German Archaeological Institute in Cairo] that those air shafts—they were still called air shafts at the time—should be investigated more closely because the technology to do so was now available. And they were truly the last aspects of the Great Pyramid that had not been explored yet.
>
> As a result, we installed a ventilation system in the pyramid in 1992. We examined the upper shafts with a video camera, searching for possible openings to the other shafts below. We determined in 1992 that these shafts have no openings anywhere. This left open the question of where and how the lower shafts end. This became the premise for our entire investigation.
>
> We named the new project UPUAUT-2, a name which requires some explanation of course. It was Professor Stadelmann's idea to call the robot UPUAUT. UPUAUT is an ancient Egyptian god, whose name roughly translates to "opener of paths."
>
> The whole purpose of developing UPUAUT-2 was to explore the two lower shafts.

What are these "upper" and "lower" shafts? The Great Pyramid contains three chambers. Dr. Rainer Stadelmann claims that this is true of all Egyptian pyramids. He is considered the father of the "three-chamber theory." All the tourists who clamber into the Great Pyramid are shown two of these chambers. The upper one is generously called the King's Chamber, although no mummy was found in it. The lower chamber, slightly smaller in size, is called, accordingly, the Queen's Chamber. Two tunnels extend downward, diagonally, from the upper chamber. They are generally referred to as air shafts. That is where Rudolf Gantenbrink installed his ventilation system. Tourists knew that from the whiff of fresh air that filtered into the King's Chamber—at least for a brief time. The ventilation system is no longer operational. It is not Rudolf Gantenbrink who is to blame, however, but rather the guards in the pyramid, who, for unknown reasons, keep forgetting to turn on the electricity.

The lower chamber is slightly smaller than the King's Chamber. I have already described its measurements in another part of this book. There are two tunnels extending away from this chamber as well: one due south, the other north. Obviously, the tunnel openings face each other. They are found at the same level as the entrance tunnel into the chamber. Rudolf Gantenbrink's robot explored the southern shaft. The third chamber is the "unfinished chamber" under the pyramid.

What do the experts tell us about the shafts in the Queen's Chamber? They can't quite agree. Some think they are tunnels for the souls. Others call them "model corridors," "air shafts," or "air shaft openings."[2] These last two suggestions are obviously flawed, because the "air shafts" were not opened up until the last century. In 1872, W. Dixon, a Briton, was knocking on all the walls in the chamber, hoping to locate secret chambers. When the knock became more hollow, Mr. Dixon grabbed his pick. He discovered the "air shafts," hidden behind a few inches of rock. Both shafts are square with a side length of twenty centimeters. Two things are clear. For one thing, they are not air shafts. If they were, they would have had to connect directly with the chamber, which was not the case until about 120 years ago. Moreover, the shafts must have been part of the pyramid plans from the very beginning. It would have been impossible to chisel or scrape them out of the pyramid rock during or after construction. Go ahead and measure twenty centimeters: It's too small for a child to fit through. And another thing: The two tunnels extending from the Queen's Chamber do not ascend diagonally, as they do in the King's Chamber. They first run horizontally into the pyramid, before starting their climb at an angle of exactly 39°, 36", and 28'. Most Egyptologists agreed that the shafts dead-end after a short while.[3] That was before Rudolf Gantenbrink's robot obliterated the experts' opinions.

The Opener of Paths

March 22, 1993, was as hot a day on the plain of Giza as any other, and the humidity inside the pyramid was just as high as usual. Inside the Queen's Chamber, Rudolf Gantenbrink had built himself a makeshift table out of two sawhorses and boards. The table held the command center for his remote control and a monitor which delivered crystal-clear images from the robot's camera. A video recorder gathered all the pictures on tape. An assistant gently guided the extra-thin and extra-light cable into the shaft. An Egyptologist from the Egyptian Antiquities Service was observing the monitor with increasing astonishment while Rudolf Gantenbrink, with intense concentration, operated the controls of the robot. The whole team was under much pressure, because the Egyptian Antiquities Service was planning to cancel the project on this very day. The travel agencies were complaining because tourists were not allowed to enter the Great Pyramid. The Antiquities Service was losing revenue. Admission to the Great Pyramid, after all, is not free.

Inch by inch, Gantenbrink's miniature robot climbed up the steep shaft. The headlights mounted on the front of the robot illuminated scenes which no human eyes had seen in at least forty-five hundred years—since Khufu, the (presumed) builder of the pyramid, had ruled from 2551 to 2528 B.C.

Laboriously, the robot toiled on, past smoothly polished walls, over small piles of sand and pieces of debris that had fallen from the ceiling. Finally, sixty meters deep into the bowels of the pyramid, came the first surprise: On the floor, the robot located a broken piece of metal. Right behind it was another sensational discovery: The robot's camera registered a barrier, a sliding door of sorts, which blocked the whole shaft. The upper part of the door showed two small metal brackets. Part of the left one was missing.

Rudolf Gantenbrink steered his robot toward the door

and aimed a laser beam at the door's underside. The red laser beam, five millimeters in diameter, disappeared under the door, proving that it was not resting firmly on the floor. A small piece of rock was missing from the right-hand lower corner of the door. The robot's camera recorded dark dust which, presumably, had been blown through this tiny opening over the course of thousands of years. The robot's journey, however, had ended.

Michael Haase, a mathematician from Berlin, determined the exact location of the mysterious door.[4] It is located on the south side of the pyramid, approximately fifty-nine meters above the ground, between the seventy-fourth and seventy-fifth tier of blocks. If the shaft which is blocked by the door continues at the same angle, it would reach the outer wall of the pyramid about sixty-eight meters above ground. Naturally, Rufolf Gantenbrink climbed the outside of the south wall, searching for an exit hole—without success!

A Sensation Under Wraps

Finding a sixty-meter shaft inside the pyramid is quite a sensation. So is the discovery of a door blocking it. One would expect Egyptologists to appreciate Gantenbrink's effort and personal accomplishment. One would expect him to receive praise and recognition proportionate to the magnitude of his discovery. Since it is not unusual for a star or comet to be named after the astronomer who discovered it, I will refer to the new shaft as the Gantenbrink shaft, and I applaud my colleagues who do the same.

Egyptologists hold a different view, however, one that

is clouded by narrow-mindedness and envy. They claim that the shaft has long been known to exist by other experts. That is only a quarter-truth. It is true that the existence of the horizontal tunnels extending south and north from the Queen's Chamber was known. No Egyptologist, however, had any idea that one of the tunnels was at least sixty meters long. On the contrary! The shafts were called "soul tunnels, which end after a short time."[5] Besides, assumptions do not a discovery make. One can assume all kinds of things. In reality, however, the sixty-meter shaft and the door barrier were discovered by the German engineer Rudolf Gantenbrink—and that's the truth.

Gantenbrink is not out to create a stir. His main purpose is the preservation of antiquities. He would also like to liven up archaeology with some fresh impulses and to make it more attractive again through the use of new technologies. He is a hardworking, sincere tinkerer who lends his expertise and genius to a fascinating branch of science. Evidently, Egyptologists have no use for his contribution, and Gantenbrink was muzzled.

At first, Gantenbrink's discovery of the shaft was ignored. Although this sensational event took place on March 22, 1993, with the express knowledge of the experts at the German Archaeological Institute (GAI) in Cairo and the Egyptian Antiquities Service, there was nothing but silence. The public was not informed. No one was allowed to talk about it. Most likely, the discovery would never have been acknowledged at all—or at most with a brief statement—if it hadn't been for coincidence and for Rudolf Gantenbrink himself. Gantenbrink showed a copy of his phenomenal video, recorded by the robot, to some experts. Somehow the British press got wind of it and ran a brief item entitled "Portcullis Blocks Robot in Pyramid"—two weeks after the fact![6] The news item was faxed to Cairo. What was the reaction?

The German Archaeological Institute in Cairo denied the report. "That's complete rubbish," the Institute's media representative, Christel Egorov, told the news agency Reuters.[7] She maintained that the tunnel was nothing but an air shaft and that the miniature robot had been deployed to measure the humidity inside the pyramid. She also "affirmed" that there were no more chambers inside the Great Pyramid.

Do you feel like someone is pulling the wool over your eyes? Well, you're right! The archaeologists at the German Archaeological Institute knew that the official statement was not true. The robot, after all, was not equipped with any kind of humidity gauge on its excursion into the Gantenbrink shaft. But it gets even better: Professor Rainer Stadelmann, the great pooh-bah of German Egyptology and director of the German Archaeological Institute, summarily denied the possibility that a secret chamber may be concealed behind the tunnel door. In front of journalists, he declared that it is well-known that the pyramid was long ago robbed of all its treasures.[8] One of his colleagues, Dr. Günter Dreyer, himself an Egyptologist, chimed in, "There is nothing behind this door. It's all a fabrication."[9]

Before I can go into detail about how this illustrious clique of Egyptologists from Cairo discredited Rudolf Gantenbrink, I feel compelled to examine the various views concerning the interior of the pyramid. It is totally presumptuous to claim that the pyramid holds nothing besides the three well-known chambers and that nothing could possibly be concealed behind the newly discovered door. The Egyptologists from the German Archaeological Institute would certainly be justified in arguing that *no one knows* whether the mysterious door is hiding anything. Instead, they proclaim with absolute certainty that nothing can be found behind this door—an attitude which is not only dogmatic and highly unscientific but also, to use the words of the Institute, "complete rubbish."

Ancient Knowledge

A historical note: In the fourteenth century, the libraries of Cairo guarded fragments of ancient Arabic and Coptic manuscripts which the geographer and historian Taki ad-Din Ahmed ben 'Ali ben Abd al-Kadir ben Muhammad al-Maqrizi compiled in his *Khitat*. I would like to reiterate briefly some of the contents, which are discussed earlier in this book. Al-Maqrizi wrote:[10]

> After that, he [the builder] had thirty treasure vaults made from colored granite in the western pyramid; they were filled with rich treasures, with tools and stelae made from precious stones, with tools made from excellent iron, like weapons that will not rust, with glass that can be folded without breaking, with strange talismans, with the various kinds of simple and combined healing potions and with deadly venoms. In the eastern pyramid, he created representations of the various constellations and the planets, as well as pictures of the things his ancestors had created; added to that were incense sacrificed to the stars and books about the stars. One also finds there the fixed stars and that which happens

> in their eras from time to time. . . . Finally, he had the bodies of the prophets brought to the colored pyramid in coffins made from black granite; next to each prophet was a book which described his marvelous skills, his life, and his works that he had accomplished in his time. . . .

It goes without saying that Egyptologists don't put much stock in any of these ancient Arabic writings. They are convinced, beyond the shadow of a doubt, that Khufu commissioned the construction of the Great Pyramid, even if a number of convincing arguments point to a contrary finding.

The attitude of these Egyptologists is reminiscent of the famous trio of monkeys: They hear nothing, see nothing, and say nothing. To some extent, I can understand their reluctance to believe in fourteenth-century writings. But unfortunately, they don't believe in modern science either if it happens to contradict their sacred beliefs. A number of examples from the past twenty-five years illustrate this point:

In 1968-69, Dr. Luis Alvarez, the Nobel Prize–winning physicist, conducted a radiation experiment at the Khafre pyramid. Archaeologists totally ignored the perplexing data gained during this experiment.

In 1988, two French architects, Jean-Patrice Dormion and Gilles Goidin, discovered chambers with the help of electronic equipment. Their discovery had no impact on the official views held by Egyptologists. The experiment was partially sponsored by the government-owned electric company of France, and it was consequently dismissed as a publicity stunt for Electricité de France.

Two years later, a team of Japanese scientists from Waseda University in Tokyo conducted a large-scale experiment using the most state-of-the-art electronic equipment available. The response from Egyptologists? Nothing but publicity for the Japanese electronics industry.

The Egyptologists gathered at the German Archaeological Institute in Cairo ignore absolutely everything. And their colleagues in Europe and elsewhere are usually unaware of the goings-on at Giza. If we were to trust these Egyptologists, nothing would ever be studied. There would be no need: They already know it all!

In 1992, Dr. Robert M. Schoch, a geologist at the College of Basic Studies at Boston University, gathered with other scientists to analyze geological data about the sphinx. Their conclusion: The Sphinx is at least five thousand years older than had previously been thought.[11] The commonly held view is that the sphinx was built by Pharaoh Khafre. There is no solid evidence, however, that links Khafre to the construction of the sphinx. The name "Khafre," some think, is found on a stone fragment, which, however, did not come from the sphinx but from a stele of Pharaoh Thutmose IV, who ruled more than one thousand years *after* Khafre, from 1401 to 1391 B.C.

How did Dr. Schoch, as a geologist, arrive at his conclusion that the sphinx is at least five thousand years older than Khafre? Schoch's team planted several seismographic buoys in the ground. They then created artificial sound waves, which produced an image of the area underneath the surface—a procedure which is quite common in geological studies. The results were analyzed by a computer, which produced reams of drawings with the exact outline of the sphinx. It was very clear that there were traces of erosion down to a depth of 2.4 meters, except on the back side. That side had been repaired long after the original construction of the sphinx, during the reign of Pharaoh Thutmose IV, who ordered that the sand be cleared away and the sphinx repaired.

The conclusion that must be drawn from the geological data and chemical analyses is unequivocal: The erosion was caused by a prolonged period of rains—of which there weren't any during the reign of Pharaoh Khafre—and was figured to have occurred around 7000 B.C.

How did archaeologists respond to Schoch's data? They were outraged. At a conference in Boston, Mark Lehner, and archaeologist from the University of Chicago, called Schoch a "pseudo-scientist." Lehner's main argument: If the sphinx really were that old, it would mean that there would have had to have been a culture capable of constructing such a masterpiece at the time. But humans were simple hunters and gatherers back then—they could not possibly have built the sphinx. End of story!

Many people, when put on the defensive, start slinging mud as soon as they run out of sensible arguments. They are afraid to lose. That's what happened in the debate between Dr. Mark Lehner, the archaeologist, and Dr. Robert Schoch, the geologist. Lehner attacked the credibility of his fellow scientist. Why such an unfair attack? One of the sponsors of Schoch's geological study was a certain John Anthony West. And Mr. West had committed two cardinal sins. First, he was not a scientist. Second, he had already published some books in which he entertained the possibility of a civilization older than those known to mankind. This, of course, was a sacrilege in the eyes of a "true" archaeologist.

Egyptologists did not care that Dr. Robert Schoch was by no means the only geologist involved in the seismographic studies at Giza. The team also consisted of Dr. Thomas L. Dobecki, two other geologists, an architect, and an oceanographer. They, too, were convinced that the lowest portions of the sphinx clearly suggested the presence of "water channels" that only occur in rock with prolonged exposure to rain. To no avail! Dr. Schoch's geological analysis was categorically rejected, and its fate was sealed by none other than Dr. Zahi Hawass, the current director of antiquities of Giza. Dr. Hawass called the entire study and the conclusions drawn from it "American hallucinations." He maintained that there was absolutely no scientific foundation for Schoch's revised dating of the sphinx.[12]

It is plain to see the Egyptologists are not impressed by scientific data generated with scientifically proven methods if these data do not fit their view of the world. *They* determine the appropriate view, unaware that they are shooting themselves in the foot. The public is fed up with scientific views. Any science that accepts the findings of other scientific branches only if they confirm its own views does not deserve much public credit.

Physics is another precise science. Dr. W. Wölfli of the Swiss Polytechnical Institute (ETH) in Zurich is a highly regarded physicist. He perfected the initially controversial C-14 method used for dating organic materials. In cooperation with colleagues from other universities, Dr. Wölfli analyzed sixteen samples from the Great Pyramid. They included pieces of charcoal, wood, splinters, and fragments of straw and grass. The result? All the samples proved to be, on the average, about 380 years older than the dates which Egyptologists had derived from the chronology of kings. One of the samples taken from the Great Pyramid was 843 years older than it was supposed to be.[13]

In all, the physicists examined sixty-four organic samples from the "Old Kingdom" period with various dating methods, including mass spectrometry. *All the samples*, without exception, were determined to be several centuries older than Egyptologists claimed. Yet no conclusions are drawn from these data—on the contrary: The old, unsubstantiated theories are buttressed with new excuses. "Excuses?" A harsh criticism, one might think. In fact, it is almost too kind a term for the nonsense we are supposed to believe.

How to Discredit a Decent Person

The Egyptologists at the German Archaeological Institute are trying to get rid of Rudolf Gantenbrink. Why? Didn't his robot produce a monumental discovery? Did he not invest much time and over $300,000 to help archaeology gain new insights? Were his methods suspect? Not at all. Gantenbrink's work is perfect, his results readily repeatable. Was he impolite? Unfriendly? In no way. Gantenbrink is a very pleasant sort of person. Did he promulgate unscientific speculations? Again, the answer is no. Rudolf Gantenbrink was very reserved in his assessments in front of the media. As the project leader of the UPUAUT mission in the Great Pyramid, he steadfastly maintained that no one could know if there really was anything behind the newly discovered door blocking the tunnel. In no way did Gantenbrink enter into any kind of speculation about the shaft or the door.

So where in the world did he go wrong? Why are the Egyptologists of the German Archaeological Institute trying to get rid of him?

He talked to the media. Not that he in any way pursued journalists to publicize his discovery. It was the other way around: The journalists were hounding Gantenbrink after British scientists had gotten wind of his phenomenal discovery. It is a journalist's job and duty to sniff out interesting news and to investigate it. Rudolf Gantenbrink's reaction was objective, reserved, and decent. Was he supposed to lie to the media, put them off? Gantenbrink is not a politician.

In an interesting article propagated by the German press agency dpa, Jörg Fischer, a journalist, wrote on June 27, 1994:[14]

> As they have done in the past, the gigantic pyramids of Giza have again generated mysterious and mystical fantasies. . . . The debate was triggered one year ago. . . . Rudolf Gantenbrink, a robotics expert from Munich, arbitrarily took it upon himself to inform the media of the discovery, speculating that the door was hiding a burial vault. The director of the German Archaeological Institute, Professor Rainer Stadelmann, recollecting all the nonsense that has been written about this, points out that "a German tabloid has already discovered the ashes of the pharaoh and a gold treasure [inside the chamber]."

In this way, Gantenbrink is discredited by people who should and do know better. Gantenbrink never speculated that "the door was hiding a burial vault." Members of the media, who don't know any better and are expected to believe the professors, are manipulated to discredit a person and to diminish his accomplishments. Gantenbrink did not "arbitrarily" release information to the press. At no time was he employed by the German Archaeological Institute or subject to any kind of gag order. The dpa article, which was disseminated worldwide and printed by many newspapers, was intended to spread false information. People were supposed to think that Gantenbrink had published unscientific suppositions. The article went on to say that this, in turn, aggravated the Egyptian government so much that it barred any further investigation of the pyramid shafts. What an outrageous distortion of the facts!

Errant Experts

The purpose of the dpa article becomes clear in a later passage:

> The archaeologist [Professor Stadelmann] categorically denies the possibility that another chamber may exist. After examining the pictures recorded by the remote-controlled video camera and comparing them to the other three shafts in the pyramid, he is convinced that the shaft is a model corridor. According to the religious beliefs of ancient Egyptians, the opening, which rises upward from the Queen's Chamber, was intended to allow passage for the pharaoh's soul to ascend to heaven. Stadelmann believes that the black dust in front of a large rock is residue from the eroded metal brackets on the door of the model corridor.
>
> However, his prosaic theory and his repeated admonition that no human being could have crawled through the nar-

row shaft or used it to transport a sarcophagus or treasure are largely ignored. . . .

Those who do not believe this theory or refuse to accept it for other reasons must be crazy. I can imagine why this learned expert "categorically denies the possibility that another chamber may exist." After all, he is the one who invented the "three-chamber theory." It has no room for the existence of a fourth or fifth chamber. It is quite odd. The rooms known to exist inside the pyramid at this point have a combined volume of approximately two thousand cubic meters. This includes the three chambers, their access tunnels, and the Grand Gallery. The Grand Gallery alone, however, accounts for eighteen hundred of the total two thousand cubic meters. In other words: The volume of the Grand Gallery far exceeds the volume of all three chambers combined. Yet it is not considered a fourth chamber. Such an interpretation would not fit the sacred three-chamber theory.

Moreover, the Gantenbrink shaft is supposed to be a "model corridor" because ancient Egyptians believed that the pharaoh's soul ascended to heaven. Let us ponder this view of a "model corridor" for a minute. The ancient Egyptians erected the most perfect structure in the history of mankind. It consists of approximately two and a half million blocks. The amount of preplanning required must have been phenomenal. Each and every stone, every support bracket, was precisely fitted, built for eternity. Inside the pyramid the Grand Gallery was constructed. The beams and stone slabs are so perfectly honed that it is difficult to see any cracks or joints. In order to get to the Grand Gallery, however, one must crawl through the so-called ascending corridor, bent over the duck style.

No one knows why the builder of the Great Pyramid constructed such a low corridor to lead to the Grand Gallery. Yet Professor Stadelmann knows beyond the shadow of a doubt that the Gantenbrink shaft is a "model corridor," because he compared it with "the other three shafts in the pyramid." Holy Osiris! What other "model corridors" are there in the Great Pyramid that could be used for comparison? Up to this point, the other shafts were called "air shafts"!

Next, Stadelmann claims that the Gantenbrink shaft is too small for a sarcophagus or even a treasure to fit through. How, then, does he explain the fact that the King's Chamber contains a granite sarcophagus whose dimensions exceed those of the ascending corridor? If we follow Professor Stadelmann's logic, this sarcophagus cannot possibly be inside the King's Chamber because the corridor, like Gantenbrink's shaft, is too small to transport a sarcophagus or treasure through.

According to Stadelmann, the builders of ancient Egypt constructed a "model corridor" inside this marvelous monument, which has survived so many centuries. But this corridor is invisible; it does not even connect directly to the Queen's Chamber. The connecting openings were broken out of the wall 120 years ago by Mr. W. Dixon. Through this "model corridor," the soul of the pharaoh was intended to ascend to the stars—but unfortunately, no pharaoh was ever buried inside the small Queen's Chamber, and no soul could have flown to the stars. Even if a body had been laid to rest there and if the shafts had been connecting, the pharaoh's soul still could not have escaped to the stars unhindered. After all, the Gantenbrink shaft is blocked by a rock, and according to Egyptologists, there is nothing behind that rock. Poor pharaoh!

Stadelmann's "prosaic theory" and his "repeated admonition that no human being could have crawled through the narrow [Gantenbrink] shaft or used it to transport a sarcophagus or treasure" is the epitome of nonsense.

Let us take a different perspective! Let us assume that Caliph Abdullah al-Mamun, who managed to dig a tunnel into the pyramid in 827 A.D., created a different opening into the pyramid from the one which we know today. He could have found an entrance on the seventy-fourth layer of bricks and discovered a chamber. From this chamber, a small square opening, twenty centimeters wide, one could have descended into the pyramid. Later, Egyptologists would call this chamber "Mamun's Chamber." They, too, would discover the square hole and would call it a "soul corridor," a "model corridor," or even an "air shaft." Then one day Rudolf Gantenbrink appears with a small caterpillar vehicle, which he guides sixty meters down the corridor. There the robot's path is blocked by a rock. According to the experts' myopic view, the newly discovered shaft cannot be leading to a chamber because "no human being could have crawled through the narrow shaft or used it to transport a sarcophagus or treasure." Excuse me, gentlemen! The fact is that the shaft—viewed from above—does lead into the Queen's Chamber. What kind of senseless view proclaims that it couldn't be the other way around? No one with any kind of factual knowledge would ever presume that treasures or a sarcophagus were transported through a square tunnel with a width of twenty centimeters. The mysterious door in the Gantenbrink shaft *may* very well (but does not necessarily) conceal a chamber, which may have another entrance as well, just like the Queen's Chamber. The Gantenbrink shaft may lead *out of* the Queen's Chamber or *into* it, depending on one's viewpoint. The Queen's Chamber can also be reached through another, independent entrance. The Gantenbrink shaft may lead into another chamber even if the door or the stone blocking it is permanently sealed. After all, the two shafts extending from the Queen's Chamber (the southern one being the Gantenbrink shaft) were also sealed until the last century. If Mr. Dixon had not taken a pick to the walls, we would have no knowledge of either shaft and no robot would have ever climbed the Gantenbrink shaft.

In other words: If the robot had descended into the Gantenbrink shaft *from the top*, it would have been stopped in front of the Queen's Chamber, just as it was stopped having started its journey from the bottom. The all-knowing experts would have agreed that this was the end, that there *couldn't* be anything behind the blocked passage. No one would have attempted to drill through the stone door or to dissolve it with acids. And this is called science? What has happened to curiosity and the desire to explore the unknown? Instead we are told categorically that there is nothing behind the door of the Gantenbrink shaft. And anyone who does not share this view is denounced as a lunatic with ridiculous fantasies.

At the end of the shaft, Gantenbrink's robot filmed two metal brackets, which are attached directly to the

a "model corridor" sealed by a stone door, then the shaft should be completely sealed; not a breeze should have disturbed the sealed vault. Moreover, the Gantenbrink shaft *was* completely sealed off until *one hundred years ago* and thus did *not* lead directly into the Queen's Chamber. The *left* metal bracket on the door is partially broken. The black dust, however, is located in the *right* corner. Is this the work of dust fairies? If the two metal brackets had indeed rusted simultaneously and quietly for several thousand years, the black dust would have to be found at the bottom of the door, directly below the brackets. That is not the case. It protrudes from the tiny triangle, as if a very weak breeze had blown it through the opening. Even a very faint breeze would indicate a continuation of the Gantenbrink shaft on the other side of the door. Or maybe the presence of another chamber with a separate entrance. The five-millimeter-wide laser beam emitted by the UPUAUT robot disappeared *under* the door frame. No matter whether the stone is a door or a final wall, the fact is that it does not rest firmly on the ground. Shouldn't that give us something to think about? Evidently not, because the Egyptologists have decided once and for all that it is a "model corridor." Additionally, more detailed studies are totally unnecessary.

Dubious Credibility

On August 5, 1993, the director of the Egyptian Museum in Berlin, Professor Dietrich Wildung, wrote in the *Frankfurter Allgemeine* newspaper:[16]

> Certainly, Egyptologists should be grateful to the technician [Rudolf Gantenbrink]. But he is unable to resist the lure of

> the publicity-generating boulevard press and naively stumbles into the quagmire of pyramid legends and mummy myths. Soon enough, Erich von Däniken enters the scenario, interpreting the dark dust at the bottom of the stone door as an indication that the mummy of King Khufu is buried behind the door. And where there is an untouched mummy of a pharaoh, one is sure to find the immense treasure which has sparked the imagination of people since the times of Herodotus. Automatically, the gears of hobby archaeologists start turning, and the cautious experts are denounced as being out of touch, unwilling to let go of the crusty old weight of their traditional teachings.

This is the kind of ammunition Egyptologists use to fight their battles and to discredit those holding different views. It has never once occurred to me to interpret the black dust "as an indication that the mummy of King Khufu is buried behind the door." This idea was proposed by David Keys, an archaeological correspondent for the British newspaper *The Independent*. I would never suggest the presence of a pharaoh's tomb or an alleged gold treasure in the Great Pyramid, because I believe that Khufu had nothing to do with the building of the pyramid, which is generally attributed to him. I certainly don't expect to find his tomb there.

What else could be concealed behind the door blocking the Gantenbrink shaft? Presumably the same thing that would be found in all other, as yet undiscovered chambers: various writings and documents, as we have been told by fourteenth-century Arab historians. Information for the people of the future—information which, incidentally, belongs to all of mankind. The pyramid does not have any ornamental hieroglyphics because it did not belong to one particular pharaoh—it belongs to all of mankind.

In his article in *The Independent*,[17] David Keys draws attention to another curious fact: The difference in elevation between the Queen's Chamber and the King's Chamber is twenty-one and a half meters. The same elevation is found between the King's Chamber and the door at the end of the Gantenbrink shaft. Coincidence, or evidence of another chamber?

Professor Dietrich Wildung, who propagated those lies about me in the *Frankfurter Allgemeine*, is also the president of the International Association of Egyptologists. He stated in an interview with SFB Radio in Germany that Egyptologists don't harbor the illusion that their field, their knowledge, has to make everyone happy. He said that the vision of an all-pleasing, comprehensive archaeology for everyone is nonsense, completely out of tune with market realities.[18]

Since this is the prevalent opinion among the leaders of the Egyptologists community, it is not surprising that they are not interested in public opinion at all. They already know for certain that there is nothing—absolutely nothing—to be found inside the Great Pyramid. Why should they bother listening to non-Egyptologists?

Why does the public continue to finance this exclusive clan of elitist buddies? Like the kings they supposedly study, they never give any kind of accounting to their subjects.

Now, the experts from the German Archaeological Institute would like to examine the northern shaft extending from the Queen's Chamber. Rudolf Gantenbrink agrees with them. I am more inclined to ask, however, why we shouldn't first finish the homework that we have started. Various propositions and suggestions have been submitted for opening, drilling, or even dissolving the stone door. How come the opinion, cooperation, and technical expertise of Rudolf Gantenbrink are no longer desired? How can scientists, who are generally a reasonable and humorous bunch, react with such arrogance and envy?

I have a feeling that there is more than envy involved here. The representatives of exalted Egyptology are insulted, wounded in their pride, because a nonarchaeologist made an unexpected discovery. They are furious because Gantenbrink talked to the media. Do spoiled little children ever grow up? Or could it be that they are trying to hide what's behind the door? Maybe they are trying to avoid losing face and plan to examine in secrecy what the stone door has for thousands of years concealed.

It is pointless to get angry. The fact is that the scientists in charge in Egypt do not want publicity, unless it is well-controlled publicity of their own creation. No journalists or neutral observers are to be present when the Gantenbrink shaft is explored and the mysterious door is opened or a hole drilled in it. There will be no TV cameras broadcasting pictures to the whole world or showing the walls of the shaft in every detail. Egyptologists do not want any experts from other sciences to validate the metal analyses of the door brackets.

All this childish furtiveness, supposedly, is designed to allow Egyptologists to work in peace and quiet. I empathize with this notion, but this time we are not dealing with some insignificant tomb, but rather with the Great

Pyramid, which has fascinated mankind for thousands of years. We are dealing with the most gigantic structure this planet has ever seen, one of the seven wonders of the world, a monument which has been the topic of legends and writings for thousands of years. Egyptology is missing out on a great opportunity to showcase its proper and scientifically flawless proceedings to the whole world. Here, Egyptologists have the chance to separate fact from fiction, once and for all, under the watchful eyes of all those lunatics and speculators who suspect mysteries and conspiracies at every turn.

Could it be that they are afraid there might be something behind the Gantenbrink shaft after all? Is this the real reason behind the Egyptologists' childish secretiveness? Earlier archaeologists were not that particular; journalists were present at the opening of the tombs of Tutankhamen and Sekhemkhet. Nowadays, we have media networks and live broadcasts, which could transmit the images from Gantenbrink's robot in the Great Pyramid simultaneously to homes and auditoriums worldwide. It would not be necessary for a horde of journalists to squeeze into the Queen's Chamber; the experts would be left to do their job peacefully and conscientiously. But what we need are *live* pictures taken as the shaft is being explored, not pictures that are broadcast days, weeks, or months after the fact, neatly cut and accompanied by elaborate captions.

Just imagine that the United States had kept secret a monumental event such as the first moon landing and that several weeks later NASA released censored pictures to the public. Everyone would have been outraged, and with good reason. What are you trying to hide from us? Why aren't we being informed from the very beginning? Why should taxpayers support organizations which obviously don't respect them enough to tell the truth?

The Egyptologists from the German Archaeological Institute and the Antiquities Service are afraid of publicity. Those who shun publicity and try to shroud their activities in secrecy usually have something to hide. There is no other way to look at it! Those who try to conceal things end up having to lie about them later. As long as Egyptologists choose to practice this kind of cloak-and-dagger routine, the public has no reason to believe anything they have to say. They may line up a whole procession of honest men who solemnly declare that, as expected, nothing was found behind the door of the Gantenbrink shaft. The public, doubtful and critical, will not believe them. They have blown their chance.

In ancient times, Cornelius Tacitus, a Roman historian, declared, "Those who are upset by criticism admit that they deserve it."

SOURCES

CHAPTER 1

1. Mariette, Auguste. *Le Sérapéum de Memphis*. Paris: 1857. Published by Gaston Maspero, 1882.

2. Strabo. *The Geography of Strabo*. Transl. Horace Leonard Jones, Vol. 8. Cambridge, Massachusetts: Harvard University Press, 1967.

3. Wahrmund, Adolf. *Diodor von Sicilien: Geschichts-Bibliothek*. Stuttgart: 1866.

4. Mariette, Auguste, *Le Serapeum de Memphis*.

5. *Mond, Robert*. *The Bucheum*. Vol. 1. London: 1934.

6. Strabo. *The Geography of Strabo*. Transl. Horace Leonard Jones, Vol. 8.

7. Mond, Robert. *The Bucheum*. Vol. 1.

8. Ibid.

9. Herodotus. *Great Books of the Western World*. Ed. Robert Maynard Hutchins. Vol. VI: *The History of Herodotus*. Chicago: Encyclopedia Britannica, 1952.

10. The Berkeley Map of the Theban Necropolis. University of California. 1987.

11. Grieshammer, R. "Grab und Jenseitsglaube." In *Das alte Ägypten*. Ed. Arne Eggebrecht. Munich: 1984.

12. Von Däniken, Erich. *Chariots of the Gods?* New York: G. P. Putnam's Sons, 1969.

13. Ettinger, Robert C. W. *The Prospect of Immortality*. New York: 1965.

14. Leca, Ange-Pierre. *Die Mumien*. Düsseldorf: 1982.

15. Pace, M. M. *Wrapped for Eternity*. New York: 1974.

16. Hopfner, Theodor. *Der Tierkult der alten Ägypter*. Vienna: 1913.

17. Smith, H. S. *A Visit to Ancient Egypt*. Warminster: n.d.

18. Lauer, Jean Philippe. *Saqqâra, die Königsgräber von Memphis*. Bergisch Gladbach: 1977.

19. Ibid.

20. Ibid.

21. Hopfner, Theodor. *Der Tierkult der alten Ägypter*.

22. Eberhard, Otto. *Beiträge zur Geschichte der Stierkulte in Ägypten*. Leipzig: 1938.

23. Von Däniken, Erich. *Wir alle sind Kinder der Götter*. Munich: 1987.

24. Latusseck, R., and L. Kürten. "Wie man mit Milliardenaufwand ein genetisches Wörterbuch schreibt." In *Die Welt*. No. 163/1988.

25. Unger, Georg F. *Chronologie des Manetho*. Berlin: 1867.

26. Karst, Josef. *Eusebius Werke*. Vol. 5. *Die Chronik*. Trans. from Armenian. Leipzig: 1911.

27. Waddell, W. G. *Manetho*. Cambridge: n.d.

28. Karst, Josef. *Eusebius Werke*. Vol. 5. *Die Chronik*.

29. Smith, C. E. *The Evolution of the Dragon*. London: 1919.

30. Leca, Ange-Pierre. *Die Mumien*.

31. Lauer, Jean Philippe. *Saqqâra, die Königsgräber von Memphis*.

32. Harris, James E. *X-ray the Pharaohs*. London: 1973.

33. Rowe, Alan. *Discovery of the Famous Temple and Enclosure of Serapis at Alexandria*. Cairo: 1946.

CHAPTER 2

1. Spiegelberg, Wilhelm. *Die Glaubwürdigkeit von Herodots Bericht über Ägypten im Lichte der ägyptischen Denkmäler*. Heidelberg: 1926.

2. Preston, E. James, and J. Martin Geoffry. *All Possible Worlds*. New York: 1972.

3. Herodotus. *Great Books of the Western World*. Ed. Robert

Maynard Hutchins. Vol. VI: *The History of Herodotus*. Chicago: Encyclopedia Britannica, 1952.

4. Oertel, Friedrich. *Herodots Ägyptischer Logos und die Glaubwürdigkeit Herodots*. Bonn: 1970.

5. Spiegelberg, Wilhelm. *Die Glaubwürdigkeit von Herodots Bericht über Ägypten im Lichte der ägyptischen Denkmäler*.

6. Kimball, O. Armayor. *Herodotus' Autopsy of the Fayoum*. Amsterdam: 1985.

7. Beck, Hanno. *Geographie*. *Orbis Academicus*. Munich: 1973.

8. Oertel, Friedrich. *Herodots Ägyptischer Logos und die Glaubwürdigkeit Herodots*.

9. Diodorus Siculus. *Diodorus on Egypt*. Trans. Edwin Murphy. Jefferson, North Carolina: McFarland and Company, 1985.

10. Strabo. *The Geography of Strabo*. Trans. Horace Leonard Jones. Vol. 8. Cambridge, Massachusetts: Harvard University Press, 1967.

11. Gaius Plinius Secundus. *The History of the World*. Trans. Philemon Holland. New York: McGraw-Hill, 1964.

12. Vandenberg, Philipp. *Auf den Spuren unserer Vergangenheit*. Munich: 1977.

13. Diodorus Siculus. *Diodorus on Egypt*.

14. Lepsius, Richard. *Denkmäler aus Ägypten und Äthiopien*. Berlin: 1849.

15. Lepsius, Richard. *Briefe aus Ägypten, Äthiopien und der Halbinsel Sinai*. Berlin: 1852.

16. Schüssler, Karlheinz. *Die ägyptischen Pyramiden, Erforschung, Baugeschichte und Bedeutung*. Cologne: 1983.

17. Diodorus Siculus. *Diodorus on Egypt*.

18. Kimball, O. Armayor. *Herodotus' Autopsy of the Fayoum*.

19. Lepsius, Richard. *Briefe aus Ägypten, Äthiopien und der Halbinsel Sinai*.

20. Hewison, Neil R. *The Fayoum*. Cairo: 1984.

21. Petrie, W. M. Flinders. *The Labyrinth Gerzeh and Mazghunch*. London: 1912.

22. Koerner, Joseph Leo. *Die Suche nach dem Labyrinth*. Frankfurt am Main: n.d.

23. Pieper, Jan. *Die Entdeckung des Labyrinthischen*. Wiesbaden: 1987.

CHAPTER 3

1. Rauprich, Herbert. *Cheops*. Freiburg i. B.: 1982.

2. Tarhan, E. H. *Nur 4000 Jahre Kultur?* Ahlen: 1986.

3. Borchardt, Ludwig. *Gegen die Zahlenmystik an der grossen Pyramide bei Gise*. Berlin: 1922.

4. Neuburger, Albert. *Die Technik des Altertums*. Leipzig: 1919.

5. Eggebrecht, Eva. "Die Geschichte des Pharaonenreiches." In *Das alte Ägypten*. Munich: 1984.

6. Ibid.

7. Goyon, Georges. *Die Cheops Pyramide*. Bergisch Gladbach: 1979.

8. Riedl, Oskar M. *Die Maschinen des Herodot, der Pyramidenbau und seine Transportprobleme*. Vienna: n.d.

9. Borchardt, Ludwig. *Die Entstehung der Pyramiden*. Berlin: 1928.

10. Borchardt, Ludwig. *Einiges zur dritten Bauperiode der grossen Pyramide bei Gise*. Berlin: 1932. Borchardt, Ludwig. *Die Entstehung der Pyramiden*.

11. Goyon, Georges. *Die Cheops Pyramide*.

12. Herodotus. *Great Books of the Western World*. Ed. Robert Maynard Hutchins. Vol. VI: *The History of Herodotus*. Chicago: Encyclopedia Britannica, 1952.

13. Fitchen, John. *Building Construction Before Mechanization*. Cambridge, Massachusetts: MIT Press, 1986.

14. Diodorus Siculus. *Diodorus on Egypt*. Trans. Edwin Murphy. Jefferson, North Carolina: McFarland and Company, 1985.

15. Gaius Plinius Secundus. *The History of the World*. Trans. Philemon Holland. New York: McGraw-Hill, 1964.

16. Al-Maqrizi. *Das Pyramidenkapitel in Al-Makrizi's "Hitat."* Trans. Dr. Erich Graefe. Leipzig: 1911.

17. Goyon, Georges. *Die Cheops Pyramide*.

18. Riedl, Oskar M. *Die Maschinen des Herodot, der Pyramidenbau und seine Transportprobleme*.

19. Davidovits, Joseph. *Pyramid Man-Made Stone, Myth or Facts*. Florida: Barry University, 1987.

20. Klemm, D., and R. Wagner. "First Results of the Scientific

Origin Determination of Ancient Egyptian Stone Material." Second International Congress of Egyptologists. Grenoble: 1979.

21. Davidovits, Joseph. "Le calcaire des pierres des Grandes Pyramides d'Egypte serait un béton géopolymère vieux de 4600 ans." *Revue des Questions Scientifiques*. 1986.

22. Ibid.

23. "Das Haar in der Pyramide." *Die Weltwoche*. Zurich, 27 Oct. 1983.

24. Schüssler, Karlheinz. *Die ägyptischen Pyramiden, Erforschung, Baugeschichte und Bedeutung*. Cologne: 1983.

25. Lauer, Jean Philippe. *Saqqâra, die Königsgräber von Memphis*. Bergisch Gladbach: 1977.

26. Von Däniken, Erich. *Habe ich mich geirrt?* Munich: 1985.

27. Steinbauer, Friedrich. *Die Cargo-Kulte—Als religionsgeschichtliches und missionstheologisches Problem*. Erlangen: 1971. Blumrich, Joseph. *Kasskara und die sieben Welten. Weisser Bär erzählt den Erdmythos der Hopi-Indianer*. Düsseldorf: 1979.

28. Sethe, Kurt. *Die altägyptischen Pyramidentexte*. Zurich: 1960.

29. Brugsch, Heinrich. *Die Sage von der geflügelten Sonnenscheibe nach altägyptischen Quellen*. Göttingen: 1870.

30. Warburton, William. *Versuch über die Hieroglyphen der Ägypter*. Frankfurt: 1980.

31. Krassa, P., and Reinhard Habeck. *Licht für den Pharao*. Luxemburg: 1982 *and* Habeck, "Elektrizität im Altertum." *Ancient Skies*. Vol. II. 1980.

32. Habeck, Reinhard. "Licht für den Pharao." *Ancient Skies*. Vol. II. 1983.

33. Krassa, P., and R. Habeck. *Licht für den Pharao*. Luxemburg: 1982.

34. Lurker, Manfred. *Götter und Symbole der alten Ägypter*. Bern: 1974.

35. Toth, M., and G. Nielsen. *Pyramid Power*. Freiburg i.B.: 1977.

36. Kirchner, Gottfried. *Terra-X, Rätsel alter Weltkulturen*. Frankfurt: n.d.

37. "Gibt es eine Pyramidenkraft?" *Ancient Skies*. Vol. III. 1982.

38. Brunton, Paul. *Geheimnisvolles Ägypten*. Zurich: 1966.

CHAPTER 4

1. Schüssler, Karlheinz. *Die ägyptischen Pyramiden, Erforschung, Baugeschichte und Bedeutung*. Cologne: 1983.

2. Al-Maqrizi. *Das Pyramidenkapitel in Al-Makrizi's "Hitat."* Trans. Dr. Erich Graefe. Leipzig: 1911.

3. "Chefren-Pyramide, Fluch des Pharao." *Der Spiegel*. No. 33. 1969.

4. Yoshimura, Sakuji. *Non-Destructive Pyramid Investigation—by Electromagnetic Wave Method*. Tokyo: Waseda University, 1987.

5. Sitchin, Zecharia. *Stufen zum Kosmos*. Trans. into German by Ursula von Wiese. Unterägeri: 1982.

6. Sitchin, Zecharia. "Forging the Pharaoh's Name." *Ancient Skies*. US edition. Vol. VIII. No. 2. 1981. Phillips, Gene. "Members Irate over New TV Special on Pyramids." *Ancient Skies*. Vol. XV. No. 1. 1988.

7. Sitchin, Zecharia. *Stufen zum Kosmos*.

8. Ibid.

9. Kraus, Rolf. "Zum archäologischen Befund im thebanischen Königsgrab Nr. 62." *Mitteilungen der Deutschen Orientgesselschaft*. 1986.

10. Al-Maqrizi. *Das Pyramidenkapitel in Al-Makrizi's "Hitat."*

11. Tompkins, Peter. *Cheops*. Berne: 1975.

12. Von Däniken, Erich. *Beweise*. Düsseldorf: 1977. (English title of the book is: *According to the Evidence*: London, 1977.)

13. Kautzsch, Emil. *Die Apokryphen und Pseudephigraphen des Alten Testaments*. Vols. I and II. Tübingen: 1900.

14. Bonwetsch, Nath. G. *Die Bücher der Geheimnisse Henochs, das sogenannte slavische Henochbuch*. Leipzig: 1922.

15. Roth, Rudolf. "Der Mythos von den fünf Menschengeschlechtern bei Hesiod." Verzeichnis der Doktoren. Philosophische Fakultät. Tübingen: 1860.

16. Helck, Wolfgang. *Untersuchungen zu Manetho und den Ägyptischen Königslisten*. Berlin: 1956.

17. Karst, Josef. *Eusebius Werke*. Vol. 5. *Die Chronik*. Trans. from Armenian. Leipzig: 1911.

18. Böckh, August. *Manetho und die Hundssternperiode, ein Beitrag zur Geschichte der Pharaonen*. Berlin: 1845.

19. Pessl, H. V. *Das Chronologische System Manethos*. Leipzig: 1878. Dieterich, A., and R. Wünsch. *Religionsgeschichtliche Versuche und Vorarbeiten*. Vol. III. Giessen: 1907. Waddell, W. G. *Manetho*. With an English translation. London. MCMXLVIII.

20. Diodorus Siculus. *Diodorus on Egypt*. Trans. Edwin Murphy. Jefferson, North Carolina: McFarland and Company, 1985.

21. Kautzsch, Emil. *Die Apokryphen und Pseudephigraphen des Alten Testaments*. Vols I and II.

22. Herodotus. *Great Books of the Western World*. Ed. Robert Maynard Hutchins. Vol. VI: *The History of Herodotus*. Chicago: Encyclopedia Britannica, 1952.

23. Lange, Kurt. *Pyramiden, Sphinxe, Pharaonen*. Munich: n.d.

24. Ibid.

25. Lepsius, Richard. *Briefe aus Ägypten, Äthiopien und der Halbinsel Sinai*. Berlin: 1852.

26. Gaius Plinius Secundus. *The History of the World*. Trans. Philemon Holland. New York: McGraw-Hill, 1964.

27. Diodorus Siculus. *Diodorus on Egypt*.

28. Schmid, Peter. Anthropological Institute and Museum of the University of Zurich. Personal note.

CHAPTER 5

1. Sasse, Torsten. "Der Schacht des Cheops." *GRAL*. Vol. III. No. 5 (1993).

2. Goyon, Georges. *Die Cheops Pyramide*. Bergisch Gladbach: 1979.

3. Schüssler, Karlheinz. *Die ägyptischen Pyramiden, Erforschung, Baugeschichte und Bedeutung*. Cologne: 1983.

4. Haase, Michael. "Up-uaut: Der, der die Wege öffnet." *GRAL*. Vol. III. No. 5 (1993).

5. Schüssler, Karlheinz. *Die ägyptischen Pyramiden*.

6. "Portcullis Blocks Robot in Pyramid." *The Daily Telegraph*, 17 April 1993.

7. Telex, Reuters and sda, 16 April 1993.

8. "The Great Pyramid Mystery." *Mail on Saturday*, 17 April 1993.

9. "Secret Chamber May Solve Pyramid Riddle." *Times*, 17 April 1993.

10. Al-Maqrizi. *Das Pyramidenkapitel in Al-Makrizi's "Hitat."* Trans. Dr. Erich Graefe. Leipzig: 1911.

11. "Sphinx, Riddle Put to Rest?" *SCIENCE*. Vol. 255. No. 5046 (14 February 1992).

12. West, John Anthony. *Serpent in the Sky*. Wheaton, Illinois: 1993.

13. Several authors. "Radiocarbon Chronology and the Historical Calendar in Egypt." Reprinted from *Chronologies du Proche Orient*. BAR International Series 379 (1987).

14. Fischer, Jörg. "Noch immer Spekulationen um eine Geheimkammer in der Cheops-Pyramide." 515 vvvb dpa 0185. Cairo: 27 June 1994.

15. Sasse, Torsten. Interview with Professor Dr. Rainer Stadelmann. Berlin: SFB Radio, 15 June 1993.

16. Wildung, Dietrich."Pharaomarkt: Technik der Pyramidenmystik." FAZ, 5 August 1993.

17. Keys, David. "Discovery at Pyramid Was Accidental." *The Independent*, 16 April 1993.

18. Sasse, Torsten. Interview with Professor Dr. Dieter Wildung. SFB Radio. 11 November 1993.